Range Unlimited

A History of Aerial Refueling

Bill Holder & Mike Wallace

Schiffer Military History
Atglen, PA

Acknowledgments

USAF Aeronautical Systems Center History Office
USAF Aeronautical Systems Office of Information
Tommy Guttman, IAI
Bernie Lindenbaum, Air Force Research Lab (Retired)
Paul Stevens, Sargent Fletcher Co.
Ulick McEvaddy, Omega Company (Ireland)
Charles Ramey and Mike Tull, Boeing Company Communications
Chuck Clark, Parker-Hannifin

David Folger, FRL Company
Major Tom Tighe, Major David Williams, USAF
Lt. Col Martin Rollinger, USMC
Dexter Kalt, Air Refueling System Advisory Group (ARSAG)
Paul Guge, Boeing Campany
Doug Oliver, Lockheed Martin Public Affairs
Steve Justice, Lockheed Martin
Captain Kay Stewart, 141st Air Refueling Wing, Washington Air National Guard

Book Design by Ian Robertson.

Printed in China.
ISBN: 0-7643-1159-X

We are interested in hearing from authors with book ideas on military topics.

Published by Schiffer Publishing Ltd.
4880 Lower Valley Road
Atglen, PA 19310 USA
Phone: (610) 593-1777
FAX: (610) 593-2002
E-mail: Schifferbk@aol.com.
Visit our web site at: www.schifferbooks.com
Please write for a free catalog.
This book may be purchased from the publisher.
Please include $3.95 postage.
Try your bookstore first.

In Europe, Schiffer books are distributed by:
Bushwood Books
6 Marksbury Avenue
Kew Gardens
Surrey TW9 4JF
England
Phone: 44 (0)181 392-8585
FAX: 44 (0)181 392-9876
E-mail: Bushwd@aol.com.

Try your bookstore first.

Contents

Dedication

This book is dedicated to the four crewmembers of KC-135E aircraft number 59-1452, who gave the ultimate sacrifice for the service of their country when the aircraft crashed near Geilenkirchen Air Base, Germany, following the refueling of a NATO AWACS E-3A in January 1999. All four were members of the 141st Air Refueling Wing, Washington Air National Guard, and were deployed in support of NATO training operations in Europe, providing aerial refueling support throughout the European theater.

The pilot, Major David W. Fite, was a 1980 U.S. Air Force Academy graduate who served on active duty in the USAF from 1980 unti July 1989. He joined the Washington Air National Guard in 1991 and was an examiner/instructor pilot. Major Fite lived with his wife, Kay, in Bellevue, Washington.

The co-pilot, Captain Kenneth F. Thiele, was also an Air Force Academy graduate. After graduation in 1989 he served on active duty until September 1998 as a KC-135 instructor pilot. He joined the Washington Air National Guard in September 1998. He lived in Spokane, Washington, with his wife, Michelle, where they were awaiting the birth of their first child.

Major Dave Fite

Captain Ken Thiele

Major Matt Laiho

Tech Sgt Richard Visintainer

KC-135E aircraft number 59-1452

Major Matthew F. Laiho, navigator, served in the USAF from 1982 to 1989 before joining the Washington Air National Guard. He served with the 141st Air Refueling Wing in Desert Shield and Desert Storm. Major Laiho lived in Spokane, Washington, with his wife, Valerie, and sons, Eric and Evan.

Boom operator, Techanical Sergeant Richard G. Visintainer, served in the USAF from 1968 to 1972. He joined the Washington Air National Guard in 1976 as a boom operator and served until 1980, then rejoined in 1986. In his civilian life he was a truck driver. He had two children, Bridget and Rocky, a stepdaughter, Allison, and two grandchildren. He lived in Spokane, Washington.

The men and women of the 141st Air Refueling Wing are extremely proud of these men who lost their lives while demonstrating their commitment on behalf of the United States of America.

They will be missed.

Foreword

Since the beginning of powered flight, military and commercial organizations have looked for ways to extend the range and duration of flight. Increasing fuel capacity seemed to be the best way—external drop tanks, wing tip tanks, internal fuselage tanks, bomb bay tanks, and conformal fuselage fuel tanks have all been investigated. These techniques all impose weight increases, valuable fuselage space consumption, weapon systems attachment displacement, drag, fuel system complexity, and with drop tanks, the need to jettison those tanks.

Can aerial refueling eliminate those disadvantages? Yes. While added fuel and tankage have their appropriate niches and uses, when aerial refueling is employed, the disadvantages of added fuel and tankage are not incurred.

An important value of aerial refueling is the ability to trade fuel (download fuel) for increased weapons per aircraft delivery to the target. By accomplishing aerial refueling just after take-off, fuel can be supplied in flight to make up that downloaded for more weapons.

Is aerial refueling without cost? Of course it is not, and it does have its own weight and cost penalties. However, the slight cost and weight penalties attached to the receiver aircraft do not significantly impact them, but allow them to remain optimally dedicated to their basic missions.

The future for aerial refueling is growing as tankers are fitted with wing stores and remotely piloted vehicles are considered for aerial refueling. Aerial refueling may even find its way into commercial cargo aircraft and, who knows, even passenger aircraft may be aerial refueled in the future.

Dexter G, Kalt
Former Technical Specialist, Fuel and Hazards
Branch, WPAFB, Ohio,
Chairman, Aerial Refueling Systmes Advisory
Group (ARSAG)

Dexter Kalt (Holder Photo)

Foreword

Aerial refueling is perhaps the single greatest force multiplier in modern warfare throughout the world. Its most evident impact is extending the range on strike, cargo, and ferry missions. It allows aircraft with less than a full load of fuel to launch, making more room for ordnance or cargo. Less fuel at takeoff also permits safe operation from smaller runways at austere deployment locations and forward operating bases. Air refueling has helped on the other end of the mission, as well, letting damaged aircraft that were lower on fuel than planned safely recover. These capabilities that aerial refueling provide to other aircraft make an entire force more flexible and responsive to change, which is critical to success in a high tempo combat environment.

Aerial refueling in the USA became necessary in the Cold War, as the theater of operation expanded to encompass most of the Northern Hemisphere. This was the stage for the development of the U.S. Air Force Strategic Air Command's first team: the KC-135 Stratotanker and the B-52 Stratofortress. Those two aircraft were designed from the beginning to work together, and unlike other aircraft, the long-range B-52 depends on aerial refueling in order to complete almost all of its missions. This is the team that refined aerial refueling into a practical and commonplace portion of military air missions. Later, the KC-10 would bring additional capabilities.

Flying aerial refueling missions is no more than flying a precise "close tail" formation position. The receiver is responsible for staying in position as the tanker's boom operator flies the boom into the receiver aircraft's receptacle. The boom operator then monitors the receiver's position and disconnects the boom if the receiver approaches the limits of the boom's hinge. Flying as the receiver aircraft can be especially challenging in a large aircraft, since the interrelated aerodynamic effects from flying large aircraft in close vertical proximity are very complex.

Major Thomas R. Tighe, USAF,
B-52 Flight Instructor

Major Thomas Tighe (Tighe Family Photo)

Foreword

The U.S. Navy, U.S. Marine Corps, and several foreign air forces use probe and drogue refueling systems. The drogue system consists of a fuel hose on a reel and a basket containing the female connection fitting. The drogue is either mounted in a pod, carried under the tanker aircraft, or it is built into the fuselage of the tanker aircraft. Podded drogue systems are used extensively so that tanker aircraft can be reconfigured easily to perform other missions. Probes are the male connection hardware fitted on the receiver aircraft. The probes are either bolted in place or made retractable, thus reducing aircraft drag.

Aerial refueling operations commence as the receiver aircraft approaches the tanker aircraft from the left side. The receiver aircraft signals that he is ready to accept fuel, while the tanker aircraft reels out the hose and basket. The receiver, with probe extended, makes contact with the basket and then drives the basket forward a few feet to ensure connection and proper drogue reel response. The tanker crew controls fuel offload. The tanker and receiver crews communicate primarily with hand and light signals, avoiding voice communication whenever possible. When the receiver is finished, the probe is backed out of the basket and slides over to the right wing position on the tanker. From this position, the receiver will depart or assist the tanker as it retracts the drogue.

Aerial refueling at sea has saved many lives and prevented the loss of many naval aircraft. When the aircraft carrier is far from suitable divert fields, tanker aircraft are always airborne or on alert. These tanker aircraft are able to extend the flight time of airborne fighters in the event that the single, tiny landing area on the carrier becomes usable for a short time. The tankers provide the commander with a great deal of flexibility.

Lt Col Martin Rollinger, USMC
F/A-18 Pilot

Lt Col Martin Rollinger (Rollinger Family Photo)

Introduction

Ever since the first aircraft was designed, engineers have always had thoughts on how to make it go faster and further.

Going faster was a pretty straight-forward undertaking, ie more-powerful engines and cleaner aerodynamic design. But making a plane fly further is a whole other consideration. Through the years of powered flight, there have been many experiments attempting to push out a plane's range, with many efforts resulting in failure.

It isn't as simple as just adding more fuel to the airframe. Granted, in some situations that would add to the range, but it would also affect other characteristics of the aircraft, thus cutting significantly into its performance.

One of the first appreciations of the advantages of increased range occurred during the WWII Allied air war over Europe, when unprotected bombers had to go it alone over enemy territory. Because of their limited range, fighters were forced to turn back long before the bombers reached their targets.

But late in the war, the simplest of solutions came to light with jettisonable fuel tanks which could be used until enemy contact, at which time they were dropped. That technique continues today with modern fighters, such as the F-15 Eagle and F-16 Fighting Falcon, both using wing and fuselage mounted fuel tanks for certain combat scenarios.

Another technique for increasing range that received attention as early as the 1930s is the so-called Parasite technique, where a large plane carried a smaller fighter-type aircraft to a position near the target. At that time, the still-fully-fueled fighter should have the necessary fuel to accomplish its mission and return to base or the mother aircraft with fuel to spare.

A sub-set of that technique took place in the U.S. during the 1950s with several "wing coupling" experiments. These were somewhat dangerous, to say the least. The technique involved the joining—at the wingtips—of smaller planes to a mother ship, which then effectively made them a single aircraft without an appreciable reduction in the performance of the mother ship. During the transport phase of the flight, the engine on the "carried" aircraft was not operating.

At least two such coupling programs were conducted during this time period, the first being a C-47/P-47 combination, followed by a B-36, which had the capability of coupling a pair of F-84s to the tips of its massive wings.

The B-36 and F-84 were also involved in another joining experiment called the F-84 FICON program. With this program, the F-84s had the capability to hook up and be dropped from a rigging underneath the B-36, thus giving it lots of "free miles."

And, amazingly, during the 1990s there have also been some experiments involving actually towing an aircraft with a line behind a puller aircraft, not unlike the technique employed by C-47s with gliders during D-Day in WWII. During the 1950s, the Air Force also towed helicopters to the site of the mission with its blades autogyrating. The technique of hooking a helicopter to the tow line proved to be much more difficult than expected, and the program was terminated.

Wing coupling experiments showing a joining of a C-47 and P-47 fighter. (USAF Photo)

So, when all was said and done, it came down to one simple conclusion, with the following steps: accepting fuel from a transport aircraft designed from scratch to carry fuel to a predesignated point; meet the aircraft to be refueled; transfer the fuel to the receiving aircraft; and release the receiving aircraft, which then proceeded to its destination with a full load of fuel. This was the best way of increasing range.

The advantage of this technique, of course, was the fact that the receiving aircraft didn't have to carry extra fuel from take-off, thus allowing it to automatically have longer range because of lighter weight.

Other advantages of the aerial refueling concept include the elimination of the problem of denial of ground landing sites. Also, the aerial refueling equipment weight/cost penalty is very small in comparison to the increased fuel, tankage, and hardware penalty that might otherwise be imposed.

As the use of tankers starting in the 1950s has matured, the use of tankers has become an important part of modern combat operations. The tankers are almost as important as the combat aircraft themselves, since the tankers are needed to enable completion of the mission. The tanker, with its unique mission, usually gives up any offensive mission, with no air-to-air or air-to-ground capabilities.

It's interesting that when aerial refueling is mentioned, there is the immediate military connotation. And certainly, that's the way it's turned out to be. But as far back as the the 1950s, with the advent of the jet airliner, there was consideration for using aerial refueling with commercial airliners. The range of the early jet transports left much to be desired, so a number of airlines looked at the possibility with interest.

With both the United States and Great Britain showing interest in the idea, there were plans to attempt aerial flight refueling with Douglas DC-4 and DC-6A airliners. The plans called for a tanker to carry an under-fuselage drogue system that would be attached to a nose-mounted receptacle. All concerned felt it important that the passengers would be unaware of the refueling taking place and would never see the tanker.

During the 1990s, there was also some consideration of aerial refueling for space applications. The idea envisioned a rocket-powered plane, about the size of a modern jet fighter, that would be fueled with hydrogen peroxide and jet fuel.

The rocket plane would take off with enough fuel to reach 60,000 feet. Then, the plane would rendezvous with a KC-135, fill up, and re-ignite for points in space. It sounds strange, but it could well work.

Studies during the late 1990s by Lockheed have also looked into the design of a new tanker configuration to eventually replace the KC-135 and KC-130 tankers. Being considered was a design that featured a "box wing," with one pair of wings sweeping back from the nose and another pair sweeping forward from the tail.

The advantages of this proposed design placed a considerable area of wing inside an overall small size. The tanker, with its simple construction, could carry considerable fuel inside its wings. There would be drogue and probe refueling systems on each wingtip. Could this be the next-generation tanker? Only time will tell.

There have also been discussions about the use of aerial refueling in the new Unmanned Aerial Vehicle (UAV) arena. It has been mentioned that existing tankers, or even smaller aircraft, could be used to refuel these unmanned vehicles once they have reached operational altitudes. Such a capability could allow a UAV such as the Global Hawk to remain on station for days instead of hours.

But when all is said and done, one thing is for sure—aerial refueling has been proven through the years and is definitely here to stay.

How the technique will evolve in the 21st century will without a doubt be very interesting. And if history is any lesson for the future, its effectiveness will continue to play a major role in any future military confrontation.

Another wing coupling experiment for range extension with a pair of F-84s attached to the wingtips of a B-36. (USAF Photo)

USA Refueling History

Although the art of in-flight refueling might seem a modern technique, its history has lasted through eight decades, with many techniques tested both in the United States and all over the world.

The technique has evolved through many crude attempts to the amazing science it is today. Following is an examination of aerial refueling through the early years in the United States. This chapter covers the advances in in-flight refueling up until the construction of pure refueling aircraft, instead of the modification of existing models to perform that function.

The first transfer of fuel between two airborne aircraft occurred in the United States over 70 years ago, when in November 1921, the first crude aerial refueling attempt was accomplished.

An artist's impression of Wesley May performing the first aerial refueling on November 12, 1921. It occurred at 3,500 feet over Long Beach, California, when he climbed between two aircraft.

The harrowing operation involved a five gallon gasoline can strapped to the back of one Wesley May, who performed with the skill of a wingwalker, working his way out to the wing skid of a JN4. The brave May then crawled up to the engine location and poured the fluid into the fuel tank. A faltering first step, but nevertheless, a first step.

Two years later, at Rockwell Field in San Diego, a DH-4B piloted by Army aviators John Richter and Howell Smith reeled out a 50-foot hose which contained a fuel shut-off valve. The line was unfurled and acquired by the lower-flying aircraft. The technique was quickly mastered, with succeeding flights showing increasing capabilities. In 1923, a flight was accomplished by the receiving aircraft, which stayed aloft for 12 hours, 13 minutes, and flew a distance of 1200 miles. Also during that same year, Richter and Smith would accomplish another refueling that would set a world endurance record of over 37 hours.

At about this same time, a U.S. Naval Officer, Godfrey Cabot, proposed a similar technique in which large, multi-engine bombers would pick up fuel from ships loitering at appropriate locations in the Atlantic. While little or nothing came of this scheme, it highlighted the need for refueling and stimulated thoughts of strategic flight.

With the technique proving encouraging, tragedy overtook the developing technology in November 1923 when the technique was demonstrated at an air show at Kelly Field, Texas. When the line was unfurled to mate up with the receiving aircraft, it became entangled with the plane's propeller, causing engine failure, and the plane crashed, killing the pilot. It would be the first, but not the last, death to occur as the technique continued to be tested.

With the malfunction, refueling attempts were curtailed in the U.S. until 1929 when Major Carl Spaatz, with a crew of five, took off in an Air Corps Fokker monoplane. With the aid of aerial refueling, the plane broke all existing records, remaining in the air for an amazing 150 hours.

But that was just the beginning, as a number of attempts followed with the records continuing to increase in time aloft. The final record took place in a plane called the Greater St. Louis, which remained aloft for over 647

An endurance record was set in 1929 by Major Carl Spaatz in a Fokker C-2. (USAF Photo)

hours. Bet the pilots were finally glad to step on solid ground after that experience!

Another significant accomplishment occurred in 1929 when a U.S. Army C-2A was refueled almost three dozen times by a Douglas C-1. The flight would have lasted longer than 150 hours had there not been engine problems.

In 1929, the Boeing Company, a leader in American refueling experiments, carried out its earliest air-to-air refueling experiments using a Model 95 mail plane as a receiver and a Model 40-B as the tanker. In these tests, a trailing hose was extended from the tanker and the two planes were flown into position so a receiver crew member could grasp the nozzle and fit it into the fuel tank filler pipe. The receiver was known as the "Boeing Hornet Shuttle."

An interesting footnote to the history of inflight refueling took place when the Hunter brothers of Chicago conducted a family aerial refueling with a pair of second-hand Stinson SM-1s. One plane was the tanker, the other the receiver. The exploit lasted an amazing 553 hours. The record, though, would only stand for several days before it was broken by the team of Jackson and O'Brine.

In the years to follow, even with the continued promise of the refueling technique, there was very little activity in the United States to improve the technique. There was, however, a small joint project with the British in 1940 when refueling equipment was fitted to a B-24 Liberator tanker and a B-17 Flying Fortress receiver. The testing, which was carried out at Eglin Field, Florida, in 1941, was successful, but WWII put an end to any follow-on activity.

In a 1942 letter from Jimmy Doolittle (stationed at Wright Field at the time) to Major General Hap Arnold, an interesting aerial refueling concept was proposed. It went something like this:

"Two aircraft take off. The ship to be refueled lets out a couple hundred feet of cable, on the end of which is a lead weight, from the extreme tail tip fitting, through which it is later to be refueled. The tanker takes up a position below and slightly behind the ship to be refueled, and a fixed gun shoots a harpoon carrying a cable at right angles to the direction of flight in such a way as to intercept the trailing cable.

The harpoon line and cable are then pulled into the tanker as it takes position above. The hose, which is on

An Army Douglas C-1 (upper aircraft) is shown refueling a Boeing Model 95 in 1929. Reportedly, this was Boeing's first experience with a refueling operation. (USAF Photo)

A family operation took place during the same time period, setting a record at a time of 553 hours using a pair of used Stinson SM-1s (Air Force Museum Photo)

a reel in the tanker, is then secured to the end of the cable and lowered away as the cable is reeled in by the receiving unit. When the cable is reeled all the way in by the to-be-refueled aircraft, an automatic fitting on the end of the hose engages in a socket, and refueling is started."

The technique was never reported as being attempted. With all those loose cables swinging around near the two planes, it might have been hard to acquire any volunteers to attempt it!

A similar concept was proposed by the Army Air Force Materiel Division in 1943 which actually specified the type of aircraft to be used.

"The method consists of trailing a cable with a weight attached from a coupling in the tail of a B-17E bomber which would fly straight and level. The cable was contacted by firing a projectile with a cable attached from a B-24D bomber, the tanker, which flew to the right, slightly lower, and behind the B-17. When the contact was made, the bomber's cable was pulled to the tanker by means of an electrically-operated reel and the weight was removed." The rest of the procedure was very similar to the 1942 proposal above.

In a later and more bizarre program, a B-24 tanker was mated with a P-38 Lightning which was to make contact with a towed fuel tank. Not surprisingly, this technique proved unsatisfactory, and undoubtedly, very dangerous.

Boeing KB-29 Tanker
In-flight refueling activity resumed following the war, when in 1947, the Air Force Air Material Command (AMC) turned to Boeing to study air-to-air refueling

A fighter pilot's view of the rear of a KB-29 with a drogue refueling receptacle being deployed. (USAF Photo)

methods and installations. Four months later, Boeing presented the results of its studies, outlining the possibilities of installing "hose-type" refueling equipment in both B-29 and B-50 bombers.

Then, in March 1948, Boeing and the Air Force conducted "Operation Drip," which were flight tests showing feasibility of passing fuel with a hose between a pair of B-29s in flight. AMC requested Boeing study the feasibility of a new refueling system with characteristics superior to the "hose-type" refueling equipment in both B-29s and B-50s.

AMC then requested Boeing to develop such a system. At Boeing's request, AMC conducted flight tests to determine optimum flight positions between B-29s in flight for refueling purposes.

A KB-29, equipped with a flying boom refueling system, is shown refueling an F-84 fighter with another waiting for fuel. (USAF Photo)

This Boeing B-50 bomber is being refueled by a KB-29P tanker. (USAF Photo)

During this period, there was also an American association with the British Flight Refueling Company to come up with a better refueling system for fighter aircraft. The Air Force sent a pair of B-29s to England to receive the modifications. One of the planes was modified with a probe mounted above the cockpit, while the second had a reel pod under each wing tip. The program also had a pair of F-84s outfitted with wing-mounted probes.

The system was later utilized under combat conditions during the Korean conflict. F-84Es of the 116th Fighter Bomber Wing were equipped with the aforementioned modification, and nine KB-29s were fitted with the hoses on tails and wingtips.

The capability was vividly demonstrated when Colonel Harry Dorris flew five combat missions for over 14 hours without returning to base. There were six refuelings accomplished during the period. The accomplishment proved that pilots could withstand long hours in a single-engine fighter aircraft.

Carrying a KB-29M designation, 92 B-29s and B-29As were converted to flying tankers at Boeing's Wichita facility. Initially called KC-29Ks, these early tankers incorporated a single large tank in each bomb-bay, each connected to a hose that could be unreeled for hook-up and fuel transfer. About 74 B-29s were modified to enable them to receive fuel.

The location of the refueling receptacle on the B-50 bomber is shown on the top of the fuselage right behind the cockpit. (USAF Photo)

This Hayes-Boeing KB-50J tanker refuels an F-100 fighter from its rear boom installation and an F-101 from a wing-mounted drogue pod. (USAF Photo)

This KB-50 tanker is shown refueling a trio of F-100 Super Sabre fighters. (USAF Photo)

In 1948 and 1949, Boeing carried out a program of modifying the hose-type refueling systems. The improvements addressed the increasing of the rate of fuel flow and extension of the utility of the aircraft so equipped.

The importance of inflight refueling was vividly demonstrated when KB-29 tankers refueled the B-50 "Lucky Lady" in the first non-stop flight around the world. The flight ended 94 hours and one minute after takeoff, with four inflight refuelings accomplished.

It was during this time period that tankers actually became operational with the Strategic Air Command (SAC) in the form of two refueling squadrons. The KB-29s were used to refuel B-29s, B-50s, and the new B-

36 bomber. They were employed until 1959, when the pure-tanker KC-97 was introduced.

In 1949, the Air Force authorized Boeing to install a flying boom receiver in a B-50D on an expedited basis. Also authorized was the modification of five B-29s into Flying Boom tankers, and to construct 35 additional kits for modification of B-29s into tankers. The first production flying boom tanker, a KB-29P, was delivered to the Air Force in March 1950. Only one B-29, however, was modified to serve as a fuel receiver from the KB-29P.

During 1950, an RB-45C reconnaissance bomber and an F-86 fighter were refueled by one of the new KB-29P tankers with the flying boom equipment. The

The KB-50 tanker was later modified with J47 jet pods to make it more compatible with high performance fighters. Here, a pair of F-101 fighters move up for fuel. (USAF Photo)

The KB-50 tankers were fitted with a number of spotlights for night refueling operations. (USAF Photo)

View of a Boeing KB-50 tanker at the Boeing facility. (USAF Photo)

Air Force Tactical Air Command (TAC) would not receive its first dedicated tankers until 1954 when it activated the first of two squadrons of KB-29s.

In July 1952, the KB-29Ps proved their worth in Operation Fox Peter One, a mass movement of the 31st Fighter Escort Wing (F-84Gs) from Georgia to the Far East. The refuelings took place over the southwest U.S. and the Pacific Ocean.

Boeing KB-50 Tanker

In 1956, TAC began receiving the new KB-50 with probe and drogue refueling gear. In all, there were 136 B-50s modified into KB-50s. The KB-50 platform was a natural evolution from the B-29, with its more advanced configuration affording performance and payload capabilities. Fitted with three refueling hoses that reeled out from tip tanks and a tail installation, the KB-50s became an indispensable part of TAC.

To help the lumbering propeller-powered tankers keep up with the high-performance F-100s, F-101s, F-105s, and B-66s of the day, it was evident that greater tanker performance was needed. It came in the form of a Hayes Aircraft Corporation modification which would carry the KB-50J designation. The "J" in this case stood for "Jet," as a pair of J-47 turbojet engines were mounted under the wings, greatly augmenting the plane's performance.

Thus, the KB-50 became a six-engine aircraft, combining both prop and jet power like the B-36, and a common saying of the time was "Four turning, two burning."

The KB-50J conversion was substantial, with a complete internal redesign and remanufacture of the airframe, along with wing strengthening, additional crew stations, and new electrical, hydraulic, and fuel systems. It was practically a new plane!

A view of a KB-50 tanker showing the wing-mounted drogue pod in the opened position. (Boeing Photo)

Realizing that a time would come when the prop-powered tankers wouldn't be able to keep up with future fighters, a B-47 was modified into a KB-47 with a drogue set-up. (USAF Photo)

The transfer fuel was carried in four large tanks: a center section tank; two bomb-bay tanks; a rear fuselage tank; and even in the bottom of the outboard engine nacelles. In all, there were almost 40,000 pounds of fuel for transfer.

Due to their speed limitations, the KB-50Js could still not accompany fighters in missions across the Atlantic Ocean, as is now done with KC-135s and KC-10s. The mission technique was to have the tankers rendezvous with the fighters at a predesignated location, accomplish the refueling operation, and then give the fighters a heading to the next tanker rendezvous point. The tankers were stationed on the east coast of the United States, in Bermuda, and the Azores in order to make these flights possible.

Single-seat fighter refueling from KB-50s was very challenging, especially over the ocean. It was necessary for the fighters would to go on dead reckoning from one refueling point to the other, with very little extra fuel in the case of a missed rendezvous. But the system worked a vast majority of the time.

Another part of the KB-50J conversion was the fitting with a battery of spotlights for night refueling operations. Also, a rotating beacon atop the vertical stabilizer aided fighters in locating the tanker.

The KB-50Js were used effectively during the early years of the Southeast Asia conflict, supporting RF-101 and F-100 sorties in northern Laos and North Vietnam. The planes were assigned to the 421st Air Refueling Squadron at Yokota.

The Vietnam tenure of the converted bomber tankers, though, was extremely short, lasting only from August to October 1964. Even so, it was a landmark in that it included the first inflight refueling for a combat deployment and the first inflight refueling in a combat zone.

By the mid-1960s, though, the time of the KB-50s was quickly coming to an end. Even still, the tankers lasted about a decade longer than the B-50s from which they were derived.

By this time, though, SAC had enough of the new KC-135s to support both its own and TAC's needs. Suddenly, the KB-50J was considered an antique of a past era.

One of the examples of the KB-50J breed is on display at the Air Force Museum at Wright Patterson Air Force Base, Ohio.

Boeing KB-47 Jet Tanker

As early as the 1950s, it was realized that prop-powered tankers would eventually fall by the wayside as the performance of jet aircraft increased.

To that end, an experiment was accomplished using the high-performance six-jet B-47 light bomber. The testing involved a pair of early-model B-47Bs, one converted into a KB-47G hose-and-drogue tanker, where an internal hose was unreeled from underneath the fuselage, along with a KB-47F receiver aircraft. The first successful transfer of fuel between the two planes was accomplished in 1953.

The Strategic Air Command was definitely interested in the concept, but hated to lose the critically-needed B-47s for this support mission. There was also a problem to be solved with refueling the B-47s, since prop-powered tankers were unable to reach the B-47's best fuel-receiving altitude, causing the bomber to have to refuel at a lower altitude, wasting both fuel and time. Then, too, there was the B-47's tendency to stall at relatively high speeds, which caused additional concern.

During the mid-1950s, the KB-47 concept underwent a substantial test program, at the same time as the KB-50J. The KB-47, though, would fall victim to economics. With an estimated cost of $2.7 million per copy, it was deemed too high a cost, since the project was also considered somewhat of an interim solution until the advent of the pure-jet tankers. To that end, the program was officially canceled in July 1957.

This photo shows the Lucky Lady II being refueled by a KB-29 M Tanker using a looped hose system. (USAF Photo)

Foreign Refueling History

Europe's enthusiasm for aviation was very high from its beginning. For example, Parisians had a ticker tape parade for the Wright brothers shortly after their first flights, and even erected a memorial to them in the City of Lights.

Besides the public's fascination for the new flying machines, the military began to realize possibilities for the new flying machines. World War I began in 1914, and aviation technology began to evolve more quickly as engineers and pilots sought lighter weight, more power, greater maneuverability, and longer range.

There were a variety of "solutions" to the problem of extending range, but it was not until 1917 that the first documented refueling of an aerial vehicle in flight took place. In September 1917, the British Royal Navy coastal C.1 airship was under tow behind the HMS Canterbury in the North Sea; the Canterbury delivered fuel to the airship by means of a long hose and compressed air. While this was the first surface-to-air refueling, it would not be the last.

In June 1918, Stephen Kolczewsky (nationality unknown) filed for a patent for a method of picking up fuel from ground stations. Presumably, this was a kind of "snag and drag" method.

Two French military pilots performed an air-to-air refueling in December 1923 using a hose. This followed a fatal accident involving U.S. Army pilots in June that caused at least one round of research there to come to an end.

By this time, British aerial refueling research was in full swing. For example, at Farnborough in February 1924, the Royal Aircraft Establishment successfully demonstrated the transfer of liquid—they used water instead of gasoline—from one British F.2b fighter to the other by means of a hose.

In Belgium, the Aeronautical Militaire's First Aeronautical Regiment modified two DeHavilland DH.9 aircraft and set a refueled endurance record on June 2 through June 4, 1928, of 60 hours, 7 1/2 minutes.

Aerial refueling techniques were changing, but still unsophisticated in the early 1930s. For example, Britain's Royal Aircraft Establishment, in 1930, added a weighted cable to the refueling hose. An observer in the receiver aircraft would snag the cable with a hooked stick and pull it in. Since the weighted cable was easier to snag in the windstream, this technique was a small step forward.

A demonstration of aerial refueling during the early 1930s between a Handley Page W.10 (Above) and an AS5 Courier where a crew member is taking the fuel line aboard. (Handley-Page Photo)

A Handley-Page W.10-GE is shown refueling Sir Allan Cobham's Airspeed Courier in 1934. After two successful refueling operations, Cobham was forced to make a belly landing. The first W.10 crashed that same day, causing official interest in refueling to wane.

In June 1935, the "Tanker," a Westland Wallace I, and the "Receiver" aircraft, a Hawker Hart I, accomplish an inflight refueling.

One of the mainstays of British refueling operations was the Vickers Virginia. In this picture, one Virginia refuels another.

In 1934, after many trials, British Flying Officer Richard Atcherly developed a method in which a tanker aircraft lowered a cable that crossed over a line trailing from the receiver aircraft. An observer in the receiver aircraft then would reel in the line with the "captured" cable. This capture technique proved to be a precursor of more modern systems.

Sir Alan Cobham, an experienced pilot who would become very influential in the development of refueling equipment and techniques, was interested in air-to-air refueling to accomplish a non-stop flight to Australia from England. To prepare, he conducted a series of experiments in the winter of 1932-33, in Ford, Sussex, using

The refueling line, which was unfurled from lower mid-fuselage, is quite visible in this photo.

A British Harrow tanker is shown refueling a Cabot seaplane in 1939.

a Handley Page W.10 as a tanker and a DH.9 as a receiver.

Deciding upon a slightly less ambitious goal of flying to Karachi (then part of India), Cobham began his flight September 22, 1934, piloting a specially built AS.5 Courier monoplane. After two successful refueling operations using W.10 tankers, Cobham had to make a forced belly-landing. The first W.10 crashed that same day, killing the four crewmembers. Although refueling had not caused either crash, official interest in refueling research waned.

Cobham became all the more determined to improve refueling operations, and, as a result, formed Flight Refuelling Ltd. (FRL). In August 1937, after comparative evaluations of FLR and the Royal Aircraft Establishment equipment, FLR received government contracts to continue refueling experiments.

After acquiring an Armstrong Whitworth 23 and Handley Page 51, FLR began to make some progress. Since the weighted cable method proved unsuitable with the AW.23, Cobham developed the wingtip method of contact.

In this method, the outer wing leading edge of the tanker contacted a weighted line trailed by the receiver. The tanker then would turn away, causing the cable to run into a latched hook on the wingtip; the hook was attached to a hose on a reel inside the tanker's fuselage. Once the contact was made, the hose could be pulled to the receiver aircraft.

Using this method, in late 1937 and early 1938, Imperial Airways' Shorts flying boat Cambria performed refueling trials without the AW.23. The success of these trials led to FRL's guarantee of refueling service any-

This modified Lancaster bomber is serving as a tanker for a F.4 Meteor fighter during a 12-hour endurance test. This was conducted just after the war.

where in the world, and the promise of delivery of 1,000 Imperial gallons of fuel in 10 minutes.

Another refueling technique used during the mid-to-late 1930s was the "Looped Hose" method. In this method, the tanker would approach the receiver from behind and below. The receiver would trail a hauling line to which was attached a sinker weight, a "bayonet" coupler, and a grapnel. The tanker would fire a projectile attached to a light line over the hauling line and make contact with the grapnel.

The tanker then would climb above and to one side of the receiver, and crewmembers would reel in the hauling line, remove the grapnel and sinker weight, and attach a hose to the coupler. Receiver aircraft crewmembers then would reel the hose into the reception coupling. After fuel was passed, the receiver aircraft would release the hose and the tanker would pull away, breaking a weak link in the hauling line to separate the aircraft.

Similar to the looped hose method was the "Ejector Method." With this system, the receiver aircraft would hang out a weighted line and the tanker would fire a line across it. The receiver would then haul in the tanker's line connected to a hose.

Although these methods were not wholly satisfactory, they did at least demonstrate the value and feasibility of inflight refueling.

For the most part, British efforts flagged during World War II due to other priorities. Reportedly, however, dur-

Post WWII British refueling testing involved the Lancaster bomber, shown here in both tanker and receiver modes. This was not attempted during the war. (FRL Photo)

ing the war, an in-flight refueling at night took place with a Harrow and an AW.23.

During the war, FRL was asked to assist in a plan to convert 600 Lancaster bombers to tankers and 600 Lincolns to receivers as part of a British bomber force that would help the U.S. attack Japan. The codename for the project was Tiger Force. The aircraft were to operate from Burma or China, 1,500 miles from their targets in Japan. FRL was to have had everything ready by early 1945 and was given "free rein" to accomplish the tasks. The U.S. capture of an island and airfield only 400 miles from Japan in early 1945, however, caused the contract to be canceled.

In the meantime, the German Luftwaffe conducted experiments using a Heinkel He 111 converted to a tanker and entertained proposals to use six-engine Junkers Ju 390s as tankers for Ju 290s. Work stopped as Allies moved closer to Germany and the need for long-range capabilities disappeared.

Following the war, the British Air Ministry showed an interest in aerial refueling of civil airliners operating on North Atlantic routes. FRL came up with several designs by November 1945, and by 1948, offered a plan for a 51-seat airliner that could be refueled in-flight. None of these projects got off the drawing board due to the advent of long range airliners. FRL continued, though, with numerous experiments using Lancaster bombers configured to be tankers or receivers.

Undeterred, Cobham persuaded BOAC to take part in trials during the winter of 1946-47. During these tests, amounting to 43 flights, BOAC aircraft were modified as receivers. They would take off from Heathrow Airport, rendezvous with an FRL Lancaster over the English Channel, and take on 1,000 Imperial gallons of fuel. Known as the Channel Trials, they included 17 at-night refueling operations using radar, for the first time enabling the aircraft to safely converge.

Later in 1947, FRL demonstrated the feasibility of locating and "mating up" with tankers at long range. To do this, FRL used Lancaster tankers based in the Azore Islands to refuel two FRL Lancasters configured as receivers and flown by British South American Airways crews. These flights reportedly lasted up to 20 hours and covered nearly 3,500 miles. Equipment continued to be developed. For example, in the late 1940s, fuel was pumped by the tanker, rather than just gravity-fed. Although Cobham continued to believe his systems would be adapted for civil use, it would never happen.

Relations between the Soviet Union and the United States deteriorated soon after World War II, and General Doolittle realized that to attack the Soviet Union,

the United States would need something with twice the range of the B-29. Told that it might take seven years to develop such an aircraft, the U.S. Air Force decided that air-to-air refueling was the answer. Doolittle soon paid a visit to FRL.

Following negotiations with FRL, the U.S. Air Force ordered immediate delivery of three sets of equipment, followed by 100 sets, for converting B-29s to KB-29M looped hose tankers. The USAF thus became the first air force to put aerial refueling into operational service, but it would not have been possible without the FRL contributions.

The USAF also needed some form of aerial refueling for its fighters. Something new was needed, since the European looped hose system wouldn't be practical for fighters. After examining a variety of methods, officials decided that the most promising one was a probe-mounted nozzle on the receiving fighter combined with a tapered funnel housing a fuel coupling on the end of the tanker's supply hose. The funnel would act as a stable drogue, and the supply hose itself could be used to guide the probe into the fuel coupling.

FRL's Lancasters were next pressed into service, and several tests determined the optimum cone shape for the drogue. A Meteor III, on loan, was equipped with a refueling probe mounted on its nose. The first successful demonstration was made April 6, 1949. Subsequent demonstrations were made for the Royal Air Force, airlines, aircraft companies, and foreign officials. The "probe and drogue" method, as it soon came to be known, describes the system to this day.

In 1951, Cobham's efforts to interest civil airline companies in aerial refueling came to an end. Officials de-

This 1949 British test shows a modified WWII bomber converted to a probe and drogue tanker and refueling a Gloster Meteor.

termined that the public would not be impressed with any missed attempts to couple with the drogue and hose.

The Royal Air Force, looking at the successes of U.S. aerial refueling operations, pressed ahead in 1951 with tests using Lincoln and Lancaster tankers to refuel 16 of 245 Squadron's Meteor F.8s, already equipped with refueling probes. The tests were successful, but funds were too scarce for any go-ahead.

The Korean War broke out in 1950 and, under contract to the United States, FRL reconfigured eight KB-29s with probe and drogue kits mounted under the rear fuselage. This was the USAF's first use of this system.

The U.S. Navy, seeing the benefits of in-flight refueling, contracted with FR Inc. (FRL's U.S.-based company) for refueling systems to be mounted in the bomb bays of North American AJ-1 Savages. Both Savages, and later, F7U-3M Cutlasses were able to refuel probe-equipped F2H-4 Banshees, F9F-8 Cougars, and F9F-2 Panthers.

The successes of these systems led to the conversion of the Convair R3Y-2 Tradewind flying boat transport to a tanker. The R3Y-2 was fitted with two podded British Sargent-Fletcher hose and drogue units under each wing, the configuration of which enabled the aircraft to refuel as many as four fighters (demonstrated in September 1956). Engine problems forced the retirement of the flying boats.

A USAF KB-29 is shown refueling a trio of RAF Meteors demonstrating the compatibility of U.S./British refueling equipment. Also note the ability to refuel three aircraft simultaneously in late 1940s/early 1950s time period.

The Soviets, having conducted some refueling experiments in the 1930s, apparently resumed efforts in the late 1940s. Their work encompassed all known systems. In addition, in 1955, Soviet Tu-167 Badgers were noted using a version of the looped hose refueling method to transfer fuel from the starboard wingtip of the tanker to the port wingtip of the receiver.

Dynamics of Aerial Refueling

Aerial refueling systems today fall into two main categories: probe and drogue, used by nearly every armed service; and the refueling boom, used mainly by the U.S. Air Force. The two systems are sometimes used together. For example, probe and drogue systems sometimes are placed on the end of refueling booms.

As we'll see, there are combinations of these systems—the Israeli tankers, configured by BEDEK, for example, use one boom and two probe and drogue systems on the same carrier aircraft. United States Air Force KC-10A Extenders are fitted with wing-mounted refueling pods to make possible the refueling of probe-equipped U.S. Navy aircraft, as well as NATO fighters.

Demonstrated in 1949 by Flight Refuelling Limited (FRL) of Great Britain to the U.S. Air Force, the probe and drogue system is the refueling system of choice for several nations.

In this system, the receiver aircraft has a fuel nozzle on the end of a probe that extends forward from (usually) the aircraft's nose. The tanker aircraft has a fuel hose long enough to provide adequate separation between the tanker and receiver. The hose's trailing end has a funnel-shaped metal cone, the drogue, designed to remain stable in the windstream.

In a refueling operation using this system, the receiver aircraft's pilot flies his aircraft's probe into the drogue coupling. As the probe enters, valves in the drogue and the receiver's probe nozzle open to allow fuel to enter the receiver aircraft's fuel system.

In this kind of refueling operation, the tanker pilot's responsibilities simply are to fly the tanker straight, level, at a consistent speed, and to pay attention to an indicator telling him when the coupling is made and when the refueling is complete. There is no need for additional personnel in the tanker for refueling operations.

Moreover, the flexible drogue hose, when reeled in, takes up a relatively small volume and thus may be placed either in the fuselage centerline or underwing, or even on wingtips. There are tanker configurations featuring multiple drogue stations that enable the refueling of as many as three receivers.

Since these systems were small, even fighter aircraft could be configured as tankers to refuel other fighters. The concept is known as the "buddy store," and is used by the U.S. Navy and the air forces of many foreign countries to give additional fuel to fighters having to wait their turns to land on a carrier. The buddy store is a self-powered pod containing a fuel pump and hose unit that can be mounted as a pylon under a wing.

Probe and Drogue - Pilot Commentary
Lt Col Martin Rollinger

"When initially encountered, probe and drogue refueling is the second most challenging task that naval avia-

Early coupling testing between a drogue refueling basket and an A-37 fixed probe. (Kalt Collection)

Close look at the nozzle on a probe of a reciever aircraft for a drogue tanker. (Kalt Collection)

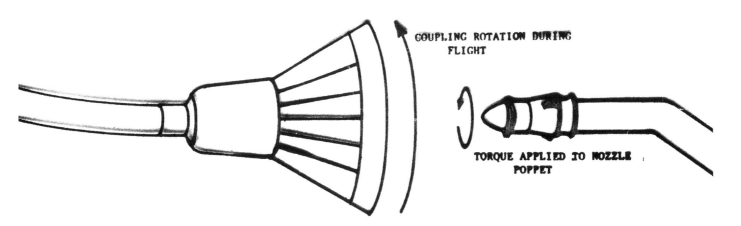

NOZZLE POPPET LOOSENING

COUPLING ROTATION DURING FLIGHT

TORQUE APPLIED TO NOZZLE POPPET

This early drawing shows the coupling rotation required to lock the two mechanisms together on a probe and drogue set-up. (Kalt Collection)

tors are required to perform. The most challenging, of course, is landing aboard the aircraft carrier. With practice, probe and drogue refueling becomes almost routine, unless the tanker is a KC-135.

Refueling an F/A-18 behind the KC-135 is particularly challenging. The KC-135 drogue is a large metal basket on a stiff nine-foot hose attached to the end of the refueling boom. There is not much room for error while attempting hook-up with the basket or once connected. What makes KC-135 refueling particularly challenging are the consequences of any errors.

The heavy basket on the KC-135 is very unforgiving when it contacts the receiver aircraft in the wrong way. The basket can literally rip the probe off the receiver aircraft. Any fuel leakage that is ingested by the turbofan engines can cause an engine to stall or, worse yet, explode.

Refueling probes are mounted on the forward fuselage of receiver aircraft near critical air data sensors. These air data sensors are easily damaged by the heavy basket. Damage these air data sensors, and the pilot has to deal with degraded handling characteristics and

Like the helicopter depicted, all aerially-refueled helicopters use the probe and drogue system. (Kalt Collection)

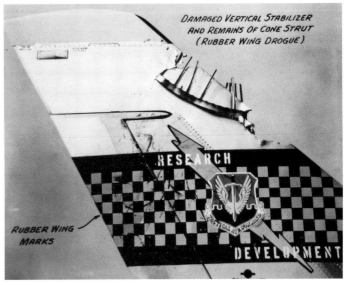

During this test of a drogue refueling system, the basket damaged the vertical stabilizer of this test aircaft. (Kalt Collection)

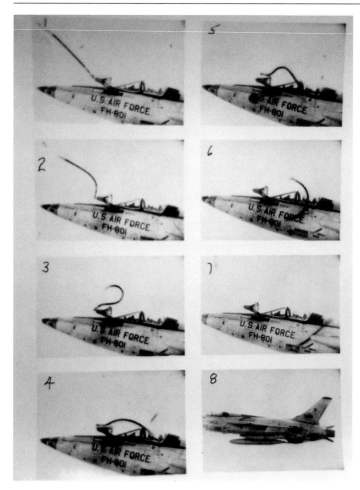

This sequence of photos shows the refueling hose broken after hook-up has been accomplished. There will be no fuel passed in this test. (Kalt Collection)

away slightly from the receiver aircraft as the approach to contact is made.

Once connected to the KC-135, the receiver aircraft needs to remain in a very small piece of sky behind the hose and basket. Remaining in proper position requires complete and total concentration for as long as it takes to receive the allotted fuel. There is no relaxing even after the connection is made. During the hook-up, while connected, and while disconnecting, the pilot is constantly aware of the hazards that he will encounter if he makes an error in aircraft positioning.

If the air is turbulent, the basket on the end of the long hose can bounce up and down plus or minus 15 feet. It's difficult to engage basket in this situation. You must wait at an intermediate position and make a stab/intercept happen as the basket comes by.

Reel response, if slack isn't taken up when the probe aircraft hits the drogue basket, is a sine wave that develops and travels up the hose and back down to the basket. The sine wave, if large enough, can sheer off the receiver aircraft's probe.

During night refueling at low altitude in the vicinity of an aircraft carrier, large numbers of receiver aircraft congregate at the tankers, and mission complexity and midair collision potential increase. Night refueling with night vision goggles adds to the complexity. The KC-

probable mission abort. Knock one of these sensors off and the pilot may have to deal with a damaged turbofan engine. This situation accounted for the loss of a Mirage 2000 over the Adriatic Sea the first day of Operation Deny Flight in 1993.

The receiver aircraft will line up with his probe about 10 feet behind the KC-135 basket. Once cleared to 'contact' the basket, the receiver will start to close with the basket at no more than one knot. While closing with the basket, the receiver pilot corrects his position (right/left and up/down) and alignment until he contacts the female connection inside the basket. Then the receiver pilot will force and maintain connection by adding power and bending the hose into a 'C' shape. If he hits the edge of the basket instead of the center, the basket will turn away from the probe, making connection impossible. At this point the receiver must carefully back away and re-attempt contact. The procedure is further complicated by turbulence, tanker turns, and the receiver aircraft bow wave that will cause the basket to move

The KC-135 was the first pure jet transport to be equipped with a boom refueling system. (USAF Photo)

The F-100 carries its probe in a unique location under the engine intake. (USAF Photo)

This Saudi F-5 is shown with its probe deployed, which is to the right and in front of the cockpit. (Kalt Collection)

130 has a different basket for helicopters, and there is danger to both aircraft due to the rotating blades.

While the probe and drogue system is simple, it has a drawback: the receiver pilot has most of the responsibility, and the tanker cannot be of much help."

Refueling Boom System

In the late 1950s, the U.S. Air Force sought an alternative aerial refueling method. Under a contract, Boeing developed the refueling boom that has changed very little from its inception. In 1960, then-Chief of Staff, General Curtis LeMay, directed that the boom would be the standard Air Force system.

The system consists of a long, pivoted telescopic pipe mounted under and to the rear of the tanker's fuselage. The boom has a nozzle on the end, as well as flight control surfaces that allow the boom operator to "steer" the boom with precision into the receiver aircraft's refueling receptacle. The receptacle has a channel with a reception coupling at its base; the receptacle generally is aft of the cockpit—partly because that's where it can be fitted, and partly because boom operators do not want the receiver pilot's "help" in lining up the boom. The boom operator can maneuver the boom from 20 to 40 degrees of depression and 15 degrees to either side of the centerline.

This S-3B takes on fuel from its retractable probe located directly above the pilot's windscreen. (U.S. Navy Photo)

Note the immense size of the fixed probe on this early Soviet fighter.

A Wright Patterson AFB test of a different probe installation for the A-37. (Kalt Collection)

The boom system offers some advantages: a greater fuel flow rate can be obtained, receiver aircraft sometimes can be towed using the boom, and "reverse refueling," in which the tanker in the normal tanker position can be refueled by a tanker in the receiver position. Reverse refueling was used in the United States' raid on Libya in the mid-1980s.

The main disadvantages of the boom are: it is more complex than the probe and drogue, it requires a boom operator, and, in its basic form, it can't be used for refueling allied and U.S. Navy aircraft.

To achieve the latter capability, Air Force refueling expert, Dexter Kalt, developed the Boom-to-Drogue Adapter (BDA) kit. The BDA mated a hose and drogue to the end of a standard boom.

Seeing contractors "reinvent the wheel" each time a new aircraft type was developed, Kalt proposed the Universal Aerial Refueling Receptacle Slipway Installation, UARRSI. The UARRSI was a move toward standardization. Today, the UARRSI may be found on nearly all modern Air Force aircraft.

Kalt further developed an advanced boom and remote operator station for what was originally called ATCA, Advanced Tanker Cargo Aircraft program, which eventually became the KC-10 Extender. The advanced boom features greater flight control surface and fly-by-wire controls. They enable the boom to be longer than that of the KC-135 to allow greater separation between tanker and receiver while keeping the precise maneuvering required for aerial refueling.

The original remote operation saw the navigator in his station become the boom operator when required. He was equipped with special goggles connected to television cameras that gave him three-dimensional

One of the few fighters to carry a fixed probe makes the A-6 instantly recognizable. (U.S. Navy Photo)

A demonstration of two types of Buddy Stores. The F-84 (in front) features a boom with a hose and drogue refueling an A-4, which in turn is refueling another A-4 using a standard probe and drogue system. (U.S. Navy Photo)

The retractable probe on the F/A-18 is located on the right side of the nose just in front of the cockpit. (U.S. Navy Photo)

images of the refueling scene. Today, the KC-10 boom operator has a remote seat and views the operation on a television monitor.

Kalt proposed hose and drogue pods for underwing and wingtip placement to allow multiple, simultaneous operations. The KC-10 is so-equipped, and so are very limited numbers of the KC-135.

He considered wing-mounted booms to be his most radical proposal. Special boom pods were constructed and successfully tested, but they were never adapted for Air Force use.

Reasoning that a tanker was a very high-value target, Kalt proposed the KA-10, a tanker version of the A-

10 Thunderbolt II. The A-10 is heavily armored and able to carry a lot of ordnance. Kalt thought that a future battle scenario could include a tanker far from the battle area, and KA-10s that would shuttle fuel from the tanker to the fighters nearer the battle area. This proposal has not been adopted…yet.

Pilot Commentary - Boom Refueling
Major David Williams

"For the KC-135 crew, preparation for air refueling begins well before takeoff. Most of the aspects of the mission revolve around this. If we are notified prior to our takeoff that our planned receiver is delayed for mainte-

The boom on a KC-135 is totally deployed to pass fuel to a C-130. (USAF Photo)

Notice the slight differences in attitude to match speeds of the two aircraft. (USAF Photo)

Looking directly down the boom during a refueling of an A-37. (Kalt Collection)

In order for USAF boom tankers to be compatible with Navy, Marine, and allied aircraft, it was necessary to have a hose and drogue attached to the end of the boom. The modification was accomplished with a Boom to Drogue Adapter (BDA) Kit. (USAF Photo)

nance, then 95 percent of the time our takeoff will also be delayed. Once airborne, our airspeed and course are adjusted to arrive at a specified point at an exact planned time.

There are two primary types of rendezvous with the receiver aircraft. For an enroute rendezvous, the tanker and receiver both plan to arrive at the same point, the

Air Refueling Initial Point (ARIP), at the same time, with 1,000 feet of altitude separating the two aircraft.

For a point parallel rendezvous, the tanker arrives at the Air Refueling Control Point (ARCP) 15 to 20 minutes prior to the planned Air Refueling Control Time (ARCT). The tanker orbits in a delay pattern waiting for the receiver. Once the receiver aircraft arrives at the

Looking directly down the boom during a refueling of an A-37. (Kalt Collection)

During B-1 testing, a simulated boom is mated to a B-1 bomber. The receptacle is above and just to the rear of the cockpit. (USAF Photo)

Note the location of the refueling receptacle on a B-1A prototype. (USAF Photo)

ARIP, the tanker flies toward him offset a planned distance from the receiver's flight path. The tanker uses radar and air-to-air TACAN Distance Measuring Equipment (DME) to measure the distance between the two aircraft. Once this distance reaches a planned turn range, the tanker turns 180 degrees toward the receiver, rolling out two to three miles in front of him, with 1,000 feet of altitude separation.

Occasionally, things don't go as planned. If it appears to either aircraft that the receiver will end up in front of the tanker, an overrun is called. In an overrun, the tanker speeds up and the receiver slows down, until the tanker overtakes the receiver.

Once the rendezvous is complete, the receiver aircraft closes to the pre-contact position, approximately 30 feet behind the tanker, and slightly below. Here the

The venerable B-52G had its receptacle located just aft of the cockpit, but out of the view of the pilot. (USAF Photo)

A test F-105 was equipped with both a receptacle and probe, giving it a dual refueling capability. This is the only model so modified. (USAF Photo)

A KC-10, equipped with an AARB, is shown refueling a B-2. The AARB is the standard boom for the KC-10. (USAF Photo)

primary responsibility of ensuring a safe contact shifts to the tanker boom operator, along with the receiver pilot.

For the actual contact, the receiver pilot uses lights on the bottom of the tanker to gage his course and altitude relative to the tanker. The boom operator has controls that enable him to maneuver the boom horizontally and vertically, as well as extend or retract the boom. Once the receiver closes into contact position, the boom operator extends the boom and inserts the boom nozzle into the receptacle of the receiver.

The tanker copilot controls the fuel panel, and thus the actual fuel offload. Once the receiver is in contact, the copilot turns on the fuel pumps of the air refueling tasks. He monitors the offload until the planned fuel has been transferred, and then turns off the pumps. Although the KC-135 is only capable of offloading fuel from two of its 10 fuel tanks, fuel from all other tanks must be

A prototype C-17 is refueled by an AARB-equipped KC-10. (USAF Photo)

A Navy F/A-18 is taking on fuel from a BDA equipped KC-135 during Operation Desert Storm. (Boeing Photo)

This USAF F-101 is receiving fuel via a probe and drogue system. (USAF Photo)

Last flight of the dual capability F-105 and first flight of the dual-capability KC-135. (USAF Photo)

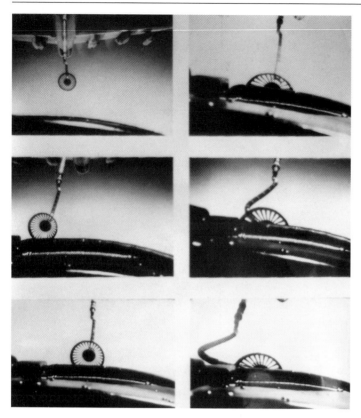

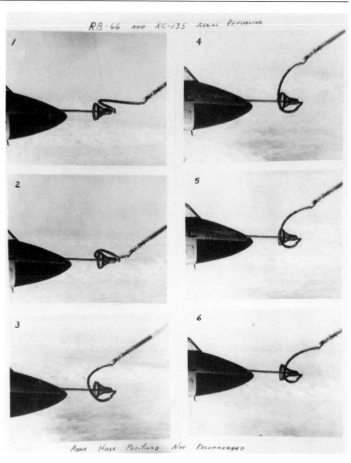

Testing of the BDA Kit was sometimes exciting. Here, the receiving is unable to hook up with the basket, which is impacting on the nose of the aircraft. The other situation shows an RB-66 having problems with a tangled hose. (Kalt Collection)

drained into the two air refueling tanks, giving the tanker the capability of offloading its entire fuel load. I have seen fuel offloads as high as 125,000 pounds (approximately 18,700 gallons) from a single tanker, when the takeoff fuel load was only 165,000 pounds.

As a tanker pilot, my primary concern during air refueling is to maintain the aircraft in a stable mode, and avoid unnecessary changes in speed, pitch, roll, and yaw. Anytime the tanker pilot makes a throttle adjustment, it requires at least three throttle adjustments on the part of the receiver pilot. The tanker pilot also tries to ensure that any turns are made as smoothly as possible.

The airspeed and altitude of the air refueling depends on the type of receiver and its mission. The KC-135 conducts refueling operations at altitudes ranging from just a few thousand feet above ground level to nearly 40,000 feet MSL, and at speeds ranging from 200 KIAS to 320 KIAS. This speed range is necessary because of the various types of aircraft routinely refueled, ranging from slow aircraft, such as the C-130 and A-10, to the fast ones, such as the B-1 and F-16.

For Navy receivers, the tanker boom is equipped with a drogue. A drogue is basically a hose that is attached to the boom with a metal basket on the end. During drogue refueling, the boom operator simply holds the boom steady. The Navy receiver aircraft has a probe on the front of his aircraft that he 'catches' in the basket. The fuel is then pumped through the hose to the receiver.

A closer view of the F-84 buddy system refueling a Navy A-4 Skyhawk. (U.S. Navy Photo)

A KC-10 refuels a pair of B-2 bombers over California. Note the location of the receptacle in the upper wing surface aft of the cockpit. (USAF Photo)

With two aircraft being as close to one another as the tanker and receiver are during air refueling, it seems that turbulence could cause problems. However, air refueling is routinely continued in conditions of light turbulence. Although the turbulence causes the aircraft to move up and down, it affects both aircraft equally, causing them to move up and down together. Moderate or severe turbulence can result in unsafe conditions.

Anytime an unsafe condition is observed by any crew member of either aircraft, such as a near collision between the two aircraft, a 'breakaway' is called. A breakaway is designed to separate the aircraft as quickly as possible. The tanker pilot advances his throttles to full power, while the boom operator moves the boom clear of the receiver. The receiver pilot reduces his power to idle and immediately dives his aircraft 1,000 feet. Once the aircraft are safely separated, the breakaway can be terminated, and the aircraft can rejoin to complete refueling.

Some of my most interesting experiences as a KC-135 pilot occurred in the Persian Gulf while deployed to Saudi Arabia in support of Desert Storm operations. We would often orbit in a refueling pattern on the edge of the Iraqi border. We would conduct pre-strike refueling with fighter aircraft heavily loaded with armaments, and then see the same aircraft return later for post-strike refueling with their armaments expended."

Summary

Besides extending range, aerial refueling gives fighters and bombers the chance to take off with maximum payloads consisting of ordnance instead of, to a great degree, fuel. Once the aircraft is airborne, it can get fuel added to it.

This USAF F-5 has a receptacle located far back on the fuselage, a practice which would be adopted for future fighters. (USAF Photo)

Close-up view of the A-7 receptacle. (Kalt Collection)

The refueling probe on this F/A-18 is shown in the deployed position. (Rollinger Photo)

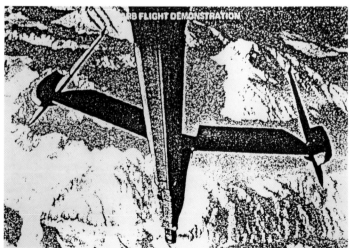

The Advanced Aerial Refueling Boom (AARB) was different from the KC-135 style in that it had a pair of vertical stabilizers attached to a horizontal wing. It gives the boom operator more control, has a fly-by-wire capability, and allows for greater separation between aircraft. (USAF Drawing)

During testing, a KC-135 was equipped with a drogue pod located under the fuselage. The technique was never employed operationally by USAF, but variations of the technique have been incoporated in foreign tankers. (USAF Photo)

Visibility is certainly a consideration in the location of the receptacle for the A-10, which used the Universal Aerial Refueling Receptacle Slipway Installation (URC). The URC is used on most modern USAF aircraft. (Kalt Collection)

UNIVERSAL AERIAL REFUELING STORE

A drawing by Air Force refueling expert Dexter Kalt depicting a wing-mounted pod system. This technique was tested on a KC-135, and it performed well. Receiver aircraft noted the lack of turbulance since there was less turbulance around the wings than the fuselage. The system was operated remotely. It was known as the Universal Aerial Refueling Store. Although it showed promise, it has yet to be installed on any operational tanker aircraft. (Kalt Collection)

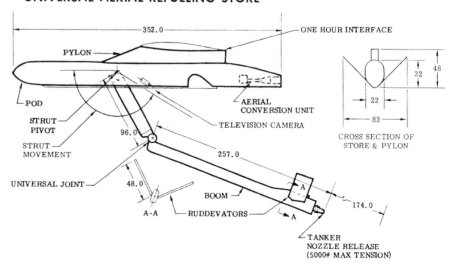

Modern United States Tankers

Background

Following the experimentation and later success of the KB-29 and KB-50 conversions, the tactical importance of tankers became extremely evident. There was even a consideration of giving the massive B-36 a tanker capability, with one model given a KB-36 designation. The single model was fitted with an FRL hose drum in the early 1950s, but the program was short-lived.

It therefore was not a surprise that new models had refueling capabilities as either a prime or joint mission. That situation was the case with the KC-97, and later, the KC-135 and KC-10.

Without tankers to this point, the U.S. Navy and Marines also became players in the refueling story with a number of systems of their own. Following is a look at each of these systems:

USAF Tankers

Boeing KC-97 Stratofreighter

The KB-29 and KB-50 tankers were large systems, but with the advent of the KC-97, size took on a new meaning. The wing and lower fuselage were basically the B-29/B-50 configuration, but that was where the similarity ended.

The plane took on a completely different look when a much larger top fuselage section was attached. It gave the plane the look of a whale, an appearance that caused it to pick up the nickname "double bubble." An interesting aspect of later versions of the KC-97 was that cargo could still be carried without removal of the refueling gear.

It is interesting that the origins of this program reached back to the early years of World War II when

three XC-97s were ordered in 1942. It was the same time that the B-29 was under development, and the Army Air Corps devoted its attention to that bomber, putting the KC-97 on the back burner.

But following the war, the interest in the KC-97 program gained momentum. The KC-97A model was first revealed in December 1950 by Boeing. That same month it was announced that a contract had been received by the contracter for construction of the KC-97E. The contract emphasized that the new plane would be a multi-mission transport, being both a cargo and a fuel carrier.

The prototype version would perform the first aerial refueling of a B-47 Stratojet bomber. It was also during this time period that the Republic F-84G was the first

A view inside the KC-97 shows the fuel storage tanks, along with still ample volume for cargo. (Air Force Museum Photo)

A close-up view of a KC-97's fuel storage tanks. (Air Force Museum Photo)

A pair of Air Force technicians view the boom operator's position on a KC-97. (Air Force Museum Photo)

Note the bubble under the KC-97 fuselage, which serves as the origination point of the boom installation. (Holder Photo)

fighter aircraft to come off the production line equipped for aerial refueling. That plane and the KC-97 would have a long relationship.

By 1956, 60 of the KC-97E tankers had been delivered to the Air Force, the first coming from Boeing in July 1951. Other versions delivered included 159 KC-97F (with improved engines) and 592 KC-97G versions. The latter version was equipped with wing-mounted tanks. A milestone was accomplished on the KC-97G production line when the 500th model was produced at Boeing's Renton, Washington, plant.

The KC-97G carried the standard flying Boom refueling system, although the aerial refueling tanks were relocated, with the model going on to serve with the Strategic Air Command (SAC). In 1957, KC-97Gs played a big part in a 24,325 mile non-stop flight around the world by three B-52s.

Finally, there was a KC-97J version which incorporated a pair of wing pylon-mounted jet engines for added speed, making the craft more compatible with the speeds of the jet-powered planes it was refueling. A number of

Even with its bulbous fuselage, the KC-97 was a graceful machine in flight. (Air Force Museum Photo)

Details of the KC-97's flying boom installation. This photo shows it in the stowed position. (Holder Photo)

The details of the boom's aerodynamic surfaces are clearly visible in this rear view of the Air Force Museum's KC-97. (Holder Photo)

The huge fuselage of the KC-97 provided huge volume for the storage of transfer fuel. (Holder Photo)

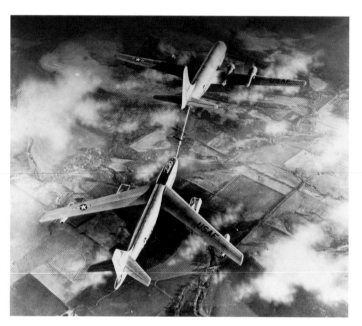

Notice the nose-attitude of the B-47 being refueled by this KC-97. (USAF Photo)

those final KC-97 variants were also used by the Israeli Air Force.

A number of fighters were modified to accept fuel from the flying boom tankers. For example, the F-84F had an automatic pilot with a receptacle located on the left wing for aerial refueling from the KC-97 and the follow-on KC-135.

Speed incompatibility, though, continued to be a problem with the KC-97 when refueling high-performance fighers and the B-47 jet bomber. In certain cases, it was necessary for the tanker to be in a shallow dive during the refueling process to acquire the necessary speed. Needless to say, that was a pretty scary deal, according to fighter and B-47 pilots that experienced it.

The mid-fuselage crease reaches the entire length of the KC-97 body. (Holder Photo)

Fred Healea of Beaver Creek, OH, was a KC-97 pilot. (Holder Photo)

Fred Healea of Beavercreek, Ohio, knows a lot about KC-97s, since he spent thousands of hours in the model as a pilot. Healea explained that the realities of the Cold War called for some precautions during the refueling of nuclear-armed bombers: "The bomber pilot had to give a code word if the tanker had to give up all its fuel. For such a case, we carried arctic survival gear for this contingency."

The strategic bomber of the day was the B-47: "We normally refueled those planes at 16,000 feet, and it was 50 minutes to altitude. We needed 200-to-210 knots of mating speed to be able to refuel jets—that was about the maximum continuous power for a prop-powered plane. Engine life was very poor with the KC-97. I've refueled B-47s, though, as high as 27,000 feet while descending—you had to descend for enough speed."

Healea recalled one B-47 mission that was nearly disastrous to him: "It was a routine operation, the B-47 was approaching to within about 400 yards lining up with the boom. All of a sudden, the boom operator started screaming that the bomber had blown up. There was no warning. Evidently, the wings had come off and the bomber exploded. If he had been closer, we'd have been lost, too."

"The design gross weight for a KC-97 was 153,000 pounds, and the standard ramp load in peacetime was 169,000 pounds. If you lost an engine on takeoff, you'd crash. If you lost one in the air, you had to dump fuel immediately. And if you turned out of traffic at more than ten degrees, you'd start to stall."

Like a number of babies waiting to feed off their mother, this bunch of F-84 Thunderjets line up for fuel behind this SAC KC-97. (Boeing Photo)

Refueling the model upon which it was based, the Boeing B-29 bomber, this KC-97 deploys its boom for the Superfortress. (Boeing Photo)

The KC-97 was still around when the Lightweight Fighter competition was held. Here, a KC-97 refuels one of the YF-17 prototypes, which would eventually become the Navy's F/A-18. (USAF Photo)

A significant photo in the history of aerial refueling. It shows the first KC-135 built, along with the final KC-97 in the background. (Boeing Photo)

Boeing KC-135 Stratotanker

Plans for the KC-135 had been in place for a number of years such that when the KC-97 was just ending production, the KC-135 model was just starting production. It was a tremendous risk on Boeing's part believing that the Air Force would continue the company line, replacing the KC-97 with the KC-135.

The 500th C-97 built just happened to be this KC-97. (Boeing Photo)

Tall tails of a squadron of KC-135s. It has been a sight that has been in place for many previous decades and will continue for a number more. (USAF Photo)

This early KC-135 carries the distintive SAC decorative band around the mid-fuselage. (USAF Photo)

An early KC-135 during a take-off. (Boeing Photo)

Interestingly enough, though, that there was some doubt at the time as to the need of this high-performance tanker. Many felt that the KC-97T could continue to fill the tanker requirement. Also, there was that $15M KC-135 pricetag for the new ship that had others concerned.

But in 1954, the Air Force—namely the Strategic Air Command—let it be known that it needed a pure-jet tanker. Suddenly, there was an enormous interest in the Boeing Model 367, the company's prototype for a commercial jet airliner. As soon as the prototype rolled out that year, the Air Force indicated it would purchase a small number of the new planes equipped with flying boom refueling systems. The first KC-135 was delivered in 1956.

Yep, not only is the KC-135 a tanker, but as this photo illustrates, it is also capable of receiving fuel in-flight. (USAF Photo)

Fan engines—PACER Craig modifications to some KC-135s—meant greater safety and cargo capacity, and allowed flight operations by only three crew members. (USAF Photo)

The KC-135 appears in just about every USAF operational and research effort. Here, a Stratotanker refuels a B-52 testbed aircraft. (USAF Photo)

A familiar sight for many years. The KC-135 is most closely associated with the B-52, which it started refueling in the 1950s and continues to refuel today. (Boeing Photo)

This test KC-135 probe was modified to do a mission it was never designed to do, that being to spray watch rearward on a trailing aircraft to produce icing conditions during its flight test program. (USAF Photo)

The C-5A transport requires a massive fill-up, but the KC-135 is up to the task. (USAF Photo)

In all, there were 732 KC-135s built, with the last being delivered in 1965. A vast majority were the KC-135A models, with the slightly-modified B version showing only 17.

The KC-135 was a heavy player in the Vietnam War, far out-distancing the numbers of Navy tankers and reaching a maximum of 195 in the 1972 time period. In addition to supporting fighter strikes, the KC-135s refueled B-52s based in Okinawa, Taiwan, and the Philippines. The KC-135 in the early 1960s was also made compatible with Navy aircraft with the addition of a hose and drogue conversion.

A measure of the level of activity of the tanker, from 1965 until the U.S. withdrawal, accounted for 195,000 sorties with some 814,000 refuelings. Almost nine billion pounds of fuel was off-loaded during the period.

There are hundreds of stories of fighters returning from their missions, damaged and almost out of fuel, that were saved by the KC-135s. Some of the fighters hooked up with hardly enough fuel in their tanks to be measured.

During the 1970s and 1980s, the F-4 Phantom was a prime customer to the KC-135. (USAF Photo)

During the flight test program for the Advanced Tactical Fighter, the KC-135 was a heavy contributor. Here, one of the two YF-22s accomplishes a refueling. (USAF Photo)

In one mission in 1966, an F-4C Phantom fighter was showing only 100 pounds of fuel as it moved in to the tanker. The fighter pilot told the boom operator to start pumping as soon as there was boom contact. It was typical of the services performed by the KC-135s through the long war.

Then, 17 years later in Desert Storm, there was the venerable KC-135 again providing its needed support to carry out the air war. In all, there were 194 KC-135s that participated in the conflict. In fact, with the exception of the 212 F-16 fighters, the KC-135 accounted for the second largest number of Coalition aircraft.

The KC-135, during Desert Storm, also refueled a large number of foreign aircraft. The Toronado (Ger-

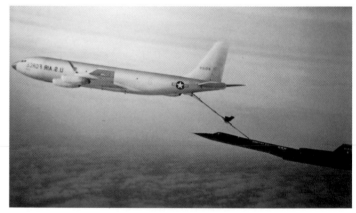

A number of changes had to be made to a select number of KC-135s refueling SR-71s because of the corrosive effects of the Blackbird's JP-10 fuel. (USAF Photo)

The YF-23 being refueled during the Advanced Tactical Fighter flight test program. (USAF Photo)

Specially-modified KC-135s were used to refuel the Mach 3 SR-71 reconnaissance aircraft. (USAF Photo)

The newest aircraft in the inventory must be compatible with one of the service's oldest aircraft, the KC-135. (USAF Photo)

many, Italy, and the UK) was a regular acceptor of fuel from the Stratotankers.

Of course, the KC-135s were also active in the second series of strikes against Iraq in the late 1990s, and still later in 1999, in the Yugoslavian conflict.

In terms of KC-135 Desert Storm statistics, the KC-135A/R/Q versions flew 9,897 sorties, accomplishing 27,390 refuelings and transferring 353 million pounds of jet fuel. For the later E version, there were 3,690 sorties with 13,391 refuelings trasferring over 164 million pounds of fuel.

A large percentage of the model are still flying in the late 1990s, and will continue to serve well into the next millennium. Of course, there have been numerous modifications to the planes through the years to increase their capabilities.

The KC-135 is different from its transport counterpart in that there would be no vision from the sides of the model, since the main fuselage is completely windowless. The fuel is carried in a dozen wing ranks in addition to nine located in the fuselage. All but a thousand gallons, which would be used by the plane itself, would be used as transfer fuel to the thirsty planes that would be coming up from behind.

The transfer is accomplished by a winged extendable flying boom which is steered by the boom operator. The maneuvering is accomplished when the boom

One particular KC-135 has been modified to perform a quite non-refueling function. Equipped with the capability to spray water through a nozzle on the end of the boom, the plane tests the capability of new aircraft to operate during icing conditions. This particular testing is being accomplished on the C-17 transport. (USAF Photo)

This artist's concept shows a KC-135 with winglets installed on its wingtips. The idea was investigated because of the improved economy that could be acquired, but the idea was never continued. (USAF Photo)

In order for the KC-135 to be compatible with Navy and USMC aircraft, it was necessary to modify the boom with a drogue extension, as can be seen here refueling a quartet of F/A-18s during Desert Storm. (USAF Photo)

A number of KC-135s are lined up on the taxiway in Saudi Arabia during Desert Storm. (Major David Williams Photo)

KC-135 crew at Lajes Air Base in the Azores during Operation Promise Hope, which was a Somalia relief effort. (Major David Williams Photo)

This photo shows a KC-135 refueling a heavily-loaded F-4 during Desert Storm. (Major David Williams Photo)

operator is lying flat on his stomach in the bottom of the rear fuselage. A row of lights on the under fuselage enables the plane being refueled to align itself for the fuel transfer.

B-52 Command Pilot, Major Tom Tighe, explained the intracacies of accomplishing a hook-up between these two large aircraft:

"The B-52 has always been a very difficult plane to refuel with the KC-135. One of the main reasons is that the B-52 is slow in roll response. There is also a vacuum effect between the two planes that sometimes tends to suck the two planes together.

The rendezvous is the most difficult aspect of the refueling operation, and requires the complete attention of the B-52 crew. You sure don't want to have your attention diverted during that operation. Bad weather also makes the operation tougher, but the two planes tend to oscillate together in bumpy conditions.

You can't see the boom when it is attached. You can hear the wind noise from the boom as it is moving into position, and then the sound of the contact. As the refueling is taking place, the B-52 is getting heavier and heavier, and as a result the plane's angle of attack (AOA) is also increasing.

But since it is necessary for the KC-135 to refuel Navy, Marine, and sometimes foreign aircraft—most of which have hose and drogue systems—compensation must be made to make the planes compatible. For the KC-135 to accomplish that job, there is a length of hose and a drogue fitted to the boom on the ground before take-off, and it remains in position for the duration of the flight. Other versions carry a fuselage-mounted Sargent-Fletcher Model FR500."

The venerable KC-135 will undoubtedly fly several decades into the 21st century, and during its already long lifetime, the model has seen a number of improvements, along with some upgrades proposed that never materialized.

A device that has found use in many commercial and business transports, i.e. wing-tip winglets, was investigated with the model with the goal of reducing drag, and therefore, possibly increasing fuel mileage. For some reason, though, the modification was not adopted.

Starting in 1975, the lower wing skins were replaced, which resulted in a considerable increase in airframe life. But probably the most significant improvement was the substitution of CFM56 turbofan engines, providing a considerable thrust increase, and certainly adding a margin of safety on take-off on hot days with a full load of fuel. A smaller number of the tankers were re-engined with TF33 engines removed from retired American Airlines 707s. These tankers would receive the aforementioned KC-135E designation.

A small number of KC-135As would be modified to carry JP-7 fuel in order to refuel the SR-71 reconnaissance aircraft. These Stratotankers would receive a KC-135Q labeling. Then, too, there was a KC-135R, or RC-135, which was classified as a reconnaissance/tanker version. KC-135F designates the French version of the tanker, and is very similar to the KC-135R configuration.

As these venerable tankers soldier on in the late 1990s, modification and maintenance work continues on the planes. Late 1990s activities included wing skin replacement, avionics system integration, and other works.

The value of the KC-135 was quickly demonstrated by a number of accomplishments when it was first introduced, the most significant taking place in 1957 when six B-52s flew non-stop from a northeastern U.S. base to Buenos Aires and returned. It marked the first time the B-52s and KC-135s had teamed up in an actual long-range tactical mission. Built as a pair, the B-52 and KC-135 have long served as the mainstay of U.S. strategic deterrence.

As a projection of future use of the KC-135, the model was part of the 4300 Provisional Bomb Wing, the first such organization, which also included B-52Gs, KC-10s, and rescue helicopters.

At presstime, the USAF announced a modification to 45 of its KC-135R tankers. To align itself with the refueling of Navy and foreign aircraft, these tankers were outfitted with wing-tip, hose and drogue refueling pods. The Multi-Point Refueling System development program was completed in 1998, with operational testing to be completed in 1999.

Thirty-three pod sets were manufactued to outfit the 45 aircraft. The pods can be moved from one tanker to another, thereby remaining mission-ready even when a particular aircraft is not. Air Force officials explained that about 7,000 working hours, or six to seven months, was required to modify each tanker, since modifications run from nose to tail, and from wing tip to wing tip. In addition, work was requireed to both the wing and fuselage fuel tanks.

Also, additional fuel controls, indicators, and circuit breakers were installed in the flight deck. Modifications to valves and a bladder cell in the fuselage were also required. Tubes, valves, and a vent system were also modified in the wing fuel system to accommodate the new system.

Even the wings were modified, strengthened to support the new fuel tubes and wire bundles, along with the hardpoints, fittings, and pylons.

Externally, the addition of the drogue system didn't appear that extensive, but looks are deceiving. But the modification will produce a much more capable refueling machine that will survive well into the 21st century.

A KC-135 simulator at Castle Air Force Base. Shown are a pilot and instructor. (David Williams Photo)

Derived directly from the DC-10 airliner, the KC-10 is shown here in flashy colors, along with the SAC emblem and stripe. (USAF Photo)

McDonnell Douglas KC-10 Extender

Even with the KC-135 serving admirably, the Air Force in the early 1970s looked to a higher-capacity tanker to augment, but certainly not replace, the Stratotanker. It was decided that a viable consideration was the use of an existing wide-body transport to fulfill the ATCA (Advanced Tanker/Cargo Aircraft) requirement. In December 1977, the announcement was made that a version of the commercial McDonnell Douglas DC-10 would be the basis for the new tanker.

The requirements placed on the new plane were strenuous, including the capability of flying global missions with several times the payload capability of the KC-135, and also the ability to provide tanker support to combat units while simultaneously carrying spares and troops.

Compared with its commercial counterpart, the KC-10—like the KC-135—sports a windowless fuselage with a large freight door. In the rear lower fuselage is the expected flying boom refueling system equipped with a fly-by-wire control system and the capability to transfer fuel at 1,500 gallons per minute from its rear boom system, which is ten feet longer than that of the KC-135. The transfer fuel is carried in seven fuel cells, each containing about 18,125 gallons.

The boom operation is refined with the KC-10 with the boom operator seated and controlling the boom position and extension maneuvers with armrest handgrip controllers. The system greatly improves operator effectiveness because of a reduction in operator fatigue.

Like the KC-135, the Extender also has the capability to refuel both with boom and drogue systems. With

An early artist's concept shows the KC-10 refueling a C-5A, a maneuver which has been accomplished in reality many times. (USAF Photo)

Unlike the KC-135, the KC-10 brought more powerful and efficient turbofan engines with it from the start. (USAF Photo)

During the KC-10 flight test program, one KC-10 refueled another, as seen in this photograph. The KC-10 can also serve as a fuel receiver with its receptacle located just above the canopy. (USAF Photo)

The KC-10 has the capability of deploying a drogue to refuel Navy aircraft like the A-4 Skyhawk shown here. (USAF Photo)

refueling drogue equipped aircraft, the pilot of the receiving aircraft must move into position such that the probe engages in the drogue at the end of a hose that is reeled out from the internal hose reel inside the Extender. The systems are located in pods on the wingtips and are frequently used to refuel Navy fighters. The Extender's drogue system is a Model FR600 by Sargent Fletcher.

The maximum loaded weight of the KC-10 is an awesome 590,000 pounds with an empty weight of almost a quarter million pounds.

Even with those weights, the Extender has a maximum speed capability of 600 miles per hour at 25,000 feet with a maximum cruising speed of 555mph. With a maximum fuel load, the plane has a range of 4,370 miles.

The KC-10 is powered by a trio of GE CF6-50C2 52,500 pound thrust turbofans. The plane's engine locations, two under the wing and one located in the tail

assembly, present no problems in performing its refueling mission. These engines are another reason for the advantages of the KC-10 with their significantly lower fuel consumption.

The Extender is also equipped with a triple inertial navigation suite. When combined with an advanced avionics system, the tanker is assured of being at the right place at the right time.

The first flight of the commercial DC-10 took place in 1970, followed by the initial KC-10 almost a decade later in July 1980. SAC received its first KC-10 in March 1981. The final order for the tanker would be 60, with 59 still in service in the late 1990s.

It also goes without saying that the KC-10 was a major player in Operation Desert Storm. Thirty of the models supported Coalition Forces along with the KC-135s. The Extenders flew 3,278 sorties, with 10,915 refuelings accomplished transferring a total of 283.6 million pounds of fuel.

Some of the KC-10s had the capability to refuel the SR-71s when they were operational. (USAF Photo)

This KC-10 shows its wingtip and fuselage-mounted drogue systems (total of three) in addition to its standard AARB boom. (USAF Photo)

KC-130s were used to refuel Jolly Green Giant rescue copters in the Vietnam War. (USAF Photo)

A pair of A-4 Skyhawks are receiving fuel from pod-mounted drogue systems on a Navy KC-130. (U.S. Navy Photo)

HC/KC-130 Hercules Tankers

The C-130 Hercules transport is considered the most versatile military transport in USAF history. Name just about any mission, and the C-130 was probably modified to accomplish it, and that certainly includes aerial refueling.

The first of numerous C-130 tanker versions was the GV-1 Sky Tanker model carrying a pair of S-F pod-mounted hose reels.

Not as well known as the high-speed refueling capabilities of the KC-135 and KC-10, the so-called HC-130 was specifically designed to refuel helicopters. A major recipiant of the HC-130's off-loading were the Jolly Green rescue helicopters on the way to and from

downed air crews during the Vietnam War. But even though the refueling of choppers was the prime mission of the model, it was still capable of refueling high-performance aircraft, as was demonstrated during the flight test program for the Navy F/A-18E/F. During the 1990s, all HC-130s are dedicated to special operations missions.

The HC-130 uses a pair of Sargent-Fletcher wing pod-mounted hose reels. There are appoximately 30 HC-130s operational during the 1990s. A number of the planes are assigned to Air Force Reserve and Air National Guard units. The Marine Corps also has a number of the planes supporting helicopter refueling operations. In 1972, a later tanker version of the Hercules

This KC-130 was used in the F/A-18 SRA program conducted by NASA. (NASA Photo)

KC-130J tanker provides significant speed, range, and fuel off-load improvements over previous KC-130 tanker variants. The first USMC KC-130J was scheduled for delivery in 2000. (Lockheed Martin Photo)

The KC-130J is capable of helicopter, as well as fixed wing aerial refueling. (Lockheed Martin Photo)

was introduced as the KC-130 Sky tanker. Like the HC-130, the model also carries two Sargent-Fletcher wing pod-mounted hose reels.

In 1989, a still-later version of the Sky Tanker was brought on line as the K/HC-130H, carrying advanced wing-mounted Sargent Fletcher Model 48-200 hose reels. A final version of the model (the K/HC-130) came on line in 1994, a model which carries a micro-processor control system.

In 1999, it was announced that five late model C-130J transports would be converted to KC-130J tankers for the USMC.

USAF Buddy Tankers
During the 1950s, the Buddy System suddenly came alive with a trio of Century Series Fighters (i.e. the F-100 Super Sabre, F-101 Voodoo, and F-105 Thunderchief) having some of their numbers modified to carry Sargent-Fletcher fuselage-mounted Buddy stores. The program was later killed by SAC Commander General Curtis LeMay, who wanted the Air Force to use the boom as the standard refueling technique. It remains that way today.

U.S. Navy/USMC Tankers

With the huge momentum the Air Force acquired in the design and acquisition of tankers, it's not surprising that the Navy and Marines both utilize the USAF tankers on a regular basis. But that is not to say that the services haven't developed tankers of their own. The Marines also have a number of C-130-based tankers to support their operations.

North American AJ-1 Savage Tanker
During the post-WWII time period, with the Air Force super active with its B-29/50 tankers and the start of the KC-97 era, the Navy came up with its first tanker, based on the AJ-1 Savage, which was carrier based. The planned purpose of the system was to increase the range of carrier-borne fighters.

The new tankers were equipped with FRL's A-12 HDU device, which was mounted in the bomb bay. Initial versions of the tanker had the hose reel located in the jet engine compartment.

Fighter aircraft modified to accept fuel from the new tanker carried a fuel probe installed in the nose. A unique connect-disconnect system assured that the fuel could be cleanly transferred between the two aircraft.

The similar AJ-2 also had a portion of its number converted to tanker configurations during the late 1950s and early 1960s. The arrangement had a hose-and-reel unit replacing the auxiliary jet engine in the tail.

A North American AJ-2 modified to a tank configuration. (NAA Photo)

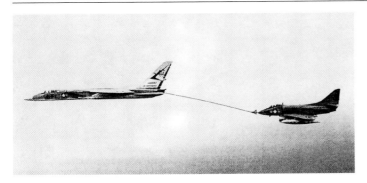

A Navy A4D moves in to refuel from an A3J Vigilante with a refueling package located in the bomb bay. (U.S. Navy Photo)

With four drogue refueling pods, the R3Y provided an attractive system for its short tenure. September 1956 saw an R3Y become the first tanker to refuel four fighters similtaneously, Grumman Cougars in this case. (U.S. Navy Photo)

Lockheed P2V Neptune Tanker

When the AJ-1 wasn't ready as soon as expected, the Navy turned to the P2V Neptune for an interim replacement. For carrier use, the model was modified with JATO racks. VC-5 was the first unit to receive the model, with additional squadrons acquiring the model later.

North American A3J Vigilante Tanker

This tanker, which was adapted from the operational Navy attack aircraft, was used for a short time. It carried a fuselage-mounted Sargent-Fletcher internal hose reel.

Chance-Vought Cutlass Tanker

With the success that the AJ-1 enjoyed, it wasn't long until a similar system was employed in the Cutlass fighter. The drogue equipped Cutlass was used to refuel Panthers, Cougars, and Banshees, all of which were equipped with fixed refueling probes. But the Navy realized that greater capability was needed, and it would not be long before that goal was met with a tanker version of the Convair R3Y-2.

Convair R3Y-3 Tradewind Tanker

It was revolutionary—that is the best way to describe the new Tradewind tanker introduced in 1954. First of all, this was a flying boat that possessed the capability to simultaneously refuel four aircraft at up to a speed of 350 miles per hour from podded units located under the wing. The tanker carried enough fuel to completely fill up eight Cougar fighters. The refueling was accomplished by a pair of podded Sargent-Fletcher Model FRL 250 hose reel pods under each wing. In late 1956, the new tanker demonstrated its capabilities by setting a new trans-Pacific speed record for a flying boat type of aircraft.

The significant power for the tanker came from four Allison T40 turbo-prop engines, each driving counter-rotating propellers. But it would be those engines, or the problems that were encountered with them, that would kill this particular project.

Douglas KA-3 Skyraider Tanker

If the KA-3 looks like it might have been an attack bomber in its past, that's exactly what it was as the A-3D. Also, in an Air Force version, it evolved as the B-66 light bomber.

In its Navy livery, the A-3D was the service's first carrier-based strategic bomber. The model served as a significant portion of America's deterrant during the Cold War. In all, some 280 of the model were built. The first

KA-3 Douglas Skyraider tanker refuels the Number One F/A-18 test model. (U.S. Navy Photo)

This Navy AJ-2 tanker is shown refueling Navy Furys. Note the lower fuselage location of the Fury's refueling probe.

This T-33 test bird, equipped with a fixed front refueling probe, is refueled by a KA-3. (U.S. Navy Photo)

flight occured in 1952, with the final production model being delivered in 1961.

But when the Skywarrier's offensive mission faded with the advent of new models, it was modified for a number of support missions, including reconnaissance, training, and the KA-3 tanker. There was also a little-known version known as the EKA-3B, a combination tanker/ECM model.

The KA-3 was equipped with a pair of pod-mounted drogue systems and was a prime USN tanker during the 1970s.

Lockheed KS-3A Tanker

The KS-3A was the follow-on to the KA-3A, based on the S-3A support aircraft, whose production ended in 1978 with 187 built. In 1980, a demonstrator version of

the KS-3A tanker version was evaluated by the Navy. The tanker was modified with a Sargent-Fletcher Model FR 400A hose reel refueling system in the rear fuselage. In addition, there were auxiliary fuel tanks added in the weapons bay. Even though the concept showed promise, it was not approved for production, and the system died.

Navy Buddy Tankers
S-3 Buddy Tanker

The S-3 Viking was adapted as a buddy refueling tanker in 1985. The system incorporates a refueling pod, carried in place of the wing-mounted external fuel tank, which contains a hose drum unit. Along with the KA-6, the S-3 buddy tanker was used extensively during the Vietnam war supporting Naval carrier operations.

This KA-3 deploys a lower-fuselage-mounted drogue system to an early F/A-18 test model. (U.S. Navy Photo)

Four F/A-18s await their turn to take on fuel from an S-3 tanker. (Photo by Lt Sjahari Pullom)

There is one amazing story of "tanker cooperation" between Navy S-3 tankers and a USAF KC-135 during the Vietnam War. It took place when a pair of S-3 tankers, themselves running extremely low on fuel, moved in to refuel from a KC-135.

Since there was only the single boom on the KC-135, the first S-3 recieved a minimal load and then backed off to allow the second Navy tanker to receive fuel. Then a dramatic situation occurred when a Navy F-8 Crusader, running out of fuel, hooked onto the second S-3 while it was still attached to the KC-135. A second F-8 then refueled from the same joined configuration. It could well have been the first dual simultaneous refueling in history.

The S-3 Tanker is a carrier-based craft completely supporting the carrier's aircraft in a loiter mode. Each Super Carrier carries an average of six S-3 tankers.

Note the location of the refueling receptacle on the S-3 tanker. (Photo by Lt Sjahari Pullom)

Turnabout is fair play. Here a USAF KC-135 refuels four S-3 Navy tankers. (Photo by Lt Sjahari Pullom)

An F/A-18 moves in for fuel from an S-3 tanker. (Photo by Lt Sjahari Pullom)

An S-3 landing on the USS John C. Stenis (CVN-74). (Photo by Lt Sjahari Pullom)

Grumman KA-6D Intruder Buddy Tanker

Interest in converting the venerable A-6 attack aircraft into a tanker conversion took place in the mid-1960s when an A-6A was converted into a tanker and first flown in May 1966. There were even production contracts placed, but eventually, they would all be canceled.

In 1970, though, the program was reborn when Grumman again developed an A-6 tanker version. It should be noted that the so-called KA-6D tankers would all be modified versions of existing A-6A and A-6E models. Externally, the tanker versions were practically identical to the attack versions. It should be noted that the later versions of the KA-6Ds deleted all weapon systems capabilities.

The KA-6D was fitted with an internal Sargent-Fletcher FR400 hose-and-refueling pod with the drogue fairing protruding under the fuselage. There was also a provision for carrying a D-704 refueling pod beneath the fuselage. The KA-6D demonstrated a capability to transfer 15,000 pounds of fuel while in a loiter mode 150 nautical miles from the carrier, or 5,000 pounds at three times that distance. Later versions of the KA-6D would increase the capabilities with 400 gallon tanks.

The KA-6D conversions were accomplished by both Grumman and the Naval Rework Facility in Norfolk, Virginia. A total of 90 conversions were accomplished. The KA-6D performed its important mission until decommissioned in 1997.

A-4/A-7 Buddy Tankers

During the 1980s, there were two additional models that got Buddy modifications, with certain numbers of the A-4 Skyhawk and the A-7 Corsair receiving the Sargent-Fletcher Model 31-301 Universal Buddy Store. Also, in 1964, an earlier version of the A-4 was equipped with a fuselage-mounted Model 31-300 Buddy Store.

Helicopter Tankers

There have been a number of investigations using helicopters as refueling tankers. The technique was demonstrated the first time in the 1950s when a pair of Marine Corps Sikorsky S-55 helicopters, rigged as a fuel tanker and reciever, accomplished the transfer. The goal was to increase the potential range and payload, along with adding more flexibility to the platform.

A USMC A-4 refuels an FJ3 with its Buddy Stores package. (Holder Collection)

The Sikorsky S-55 was used as a helicopter tanker. (USMC Photo)

A number of helicopters have been fitted with front refueling probes for drogue and probe refueling. However, there are now no helicopter tankers.

The Air Force, though, did consider the possibility of a helicopter tanker. What evolved was a "Reverse Refuel" HH-53 helicopter system. The technique involved the installation of a hose reel on the cargo deck of the chopper to accomplish the transfer of fuel. To accomplish the transfer, it was necessary to fly a KC-130 behind the HH-53. The KC-130 then pumped the fuel into the HH-53, which could then transfer it to other helicopters.

Convair 880
The Convair 880 wasn't that successful as a commercial airliner, but appeared to have the potential of a tanker. It didn't make it there, either, but there was one example assembled.

The single 880 was acquired in the early 1980s by the Navy to serve as a tanker for test operations at the Naval Air Test Center at Patuxent River. The 880 was identified as the UC-880. The plane was equipped with a single drogue along the rear lower fuselage. The plane was easily identifiable, with a mulititude of antennae and a large ventral radome. The UC-880 tanker was retired in 1994.

Future Navy/USMC Tankers
Certain models of the new F/A-18E/F have been approved to carry a Buddy refueling mission. The conversion is possible, since the E/F has two wet sections in each wing, as opposed to only one for the earlier F/A-18C/D model. The plane also has a fuselage wet section. With the refueling version, there will also be a pair of wing tanks, giving the E/F tanker five wet sections. These E/Fs will not have an offensive capability and will carry the Sargent-Fletcher 31-301 Buddy Store equipment.

Other Military Tanker Proposals

Boeing 747 Tanker Investigation
In February 1972, the Air Force contracted Boeing to conduct a flight test program using a 747 commerical aircraft modified for aerial refueling to determine the feasibility of using existing wide-body aircraft as an advanced tanker.

The tests began that same month using B-52, FB-111, F-4E, and C-135 receiver aircraft. A 747-100 was used as the simulated tanker to determine the location of a KC-135 refueling boom and the boom operator lo-

cation. The F-4 and FB-111 also investigated 747 wing tip positions to collect aerial refueling data.

Following installation of the refueling equipment, dry refueling contacts were successfully conducted at Edwards Air Force Base, California, in July 1972. It was concluded that an advanced tanker derived from such a wide-body aircraft, using existing or improved KC-135 refueling equipment, would have formating capabilities over a greater range of altitudes and speeds with substantially greater fuel off-load when compared with existing tankers.

Boeing even considered the 747 as a possible tanker, shown here in a test refueling of a B-52. (Boeing Photo)

The Boeing 707 Tanker/Transport was a drogue system refueler, shown here refueling an F-105. (Boeing Photo)

There was certainly no doubt from this flashy paint scheme that the company was really pushing this combination tanker/ transport concept. (Boeing Photo)

Of course, the follow-on to the KC-135 selected was the KC-10, effectively canceling the 747 experiment. It should be noted, though, that a small number of 747 tankers would be assembled for Middle East countries.

Douglas DC-8 Tanker Proposal

During the 1980s, it seemed that just about every airline model was considered for a tanker application. The DC-8 was examined with the use of either a boom or drogue set-up, but nothing came from the research.

Boeing 737/757 tanker Proposals

Two different Boeing 737 tanker proposals had either wing or fuselage-mounted pods, while the new Boeing 757 would carry a lower-fuselage-mounted pod.

It goes without saying that any of these arrangements would probably have worked had the Air Force decided to go in any of those directions.

Boeing 707-320 Tanker/Transport Proposal

With the success that had been enjoyed by the KC-135, Boeing decided if there were to be a new tanker, it would make good sense to use the same basic airframe where so much flight and refueling experience existed.

Hence, in 1982 the company used a modified 707 airliner and built a prototype to illustrate the concept. The model differed from the standard KC-135 in that it did not utilize the standard boom refueling system of the Stratotanker, instead using a pair of wing pod-mounted Sargent-Flectcher hose reels and a fuselage-mounted Sargent-Fletcher Model 600B internal hose reel.

Since a majority of the world's aircraft use the drogue system, and with a thought of overseas sales, Boeing equipped the new tanker with a pair of wing-mounted

pods and a centerline system that could reel out three hoses and drogues. There were also several other alternatives, with the centerline system replaced with a standard boom system that could be used in its normal manner or equipped with a hose-and-drogue set-up. The final configuration eliminated the drogue system entirely and used only a boom system, which also had the aforementioned hose-and-drogue capability.

The concept was also attractive from an economic standpoint, since there were a number of surplus airliners that could easily be converted to the new tanker configuration. In fact, the prototype was a former TWA airliner.

The prototype was detailed out beautifully in red, white, and blue, with a huge "707" lettered on the tail and "Boeing 707 Tanker-Transport" on the fuselage.

The Boeing 767 has been considered as an excellant candidate for a future tanker. This company concept shows the 767 Tanker/ Transport (front aircraft), which could evolve into a future tanker. (Boeing Photo)

Boeing future tanker designs show this blended wing concept carrying both boom and drogue systems. (Boeing Photo)

A Lockheed Martin concept of a conventional high-wing military transport converted into a multi-role tanker. (Lockheed Martin Drawing)

Possible Future U.S. Tanker Designs and Concepts

Investigations continue into the next millennium with many concept studies and flying prototypes for the development of new tankers. Following are the more important of these efforts:

Boeing 767 Tanker/Transport Proposal

The design characteristics of the relatively new Boeing 767 transport presented excellent opportunities for becoming a tanker in research work accomplished during the late 1990s. With its previous experience with 707-based tankers and research with its other airliners for the refueling mission, the company was greatly qualified to accomplish this investigation. Boeing officials explained that a commercial 767 configured for the tanker mission involved the addition of pumps and auxiliary fuel tanks, and required fuel distribution lines below the main cabin floor. The concept leaves the main cabin free for cargo or passengers, allowing for simultaneous refueling and airlift operations, or successive

Could the C-17 Globemaster II be the next Air Force tanker? Although the Boeing 767 and others have been considered, the C-17, with its load-carrying and STOL capabilities, could provide some tactical advantages. (USAF Photo)

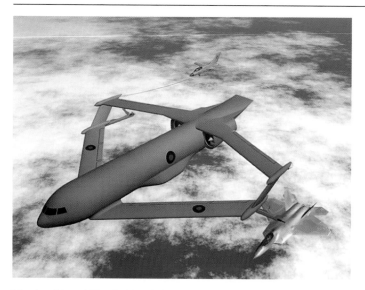

The Lockheed Martin Box-Wing tanker-transport simultaneously refueling an F-22 by refueling boom and refueling a Joint Strike Fighter by a hose and reel system. (Lockheed Martin Photo)

A horizontal plate joins the upper and lower wings. The actual refueling equipment is located on wingtip pods, the upper pods housing the hose and reel system and the lower pods the boom systems. (Lockheed Martin Photo)

sorties without time-consuming reconfiguration of any kind.

The 767 can be modified to accommodate refueling wing pods or a centerline refueling hose for probe and drogue refueling, or a combination of each. At presstime, the program was still being investigated, but it appeared very promising.

Boeing Blended Wing Tanker Concept
Boeing in the late 1990s was studying a number of advanced tanker concepts using a so-called blended wing design. Artist's concept drawings show a large wing design with both boom and drogue refueling systems illustrated.

Lockheed KC-17 Tanker
During the late 1990s, the four-jet C-17 transport was in series production and was being considered for many different missions besides just freight hauling. One such mission was, not surprisingly, that of a tanker conversion. Interest in the proposed modification came from the fact that such a tanker would be a smaller, more flexible model than the KC-135 tanker it would be replacing.

Lockheed Conventional Tanker/Transport Concepts
Lockheed has studied a conventional high-wing military transport converted into a multi-role tanker. Developed under the KC-X program, the tanker would used a fuselage-mounted flying boom and multi-point hose/drogue system.

The newest Navy fighter, the F-18E/F, will have a certain number modified as Buddy Tankers. (McDonnell Douglas Photo)

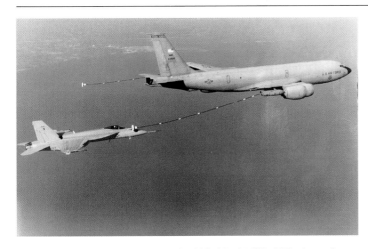

Taking of the configuration of a KC-10, 45 KC-135s have been modified with wingtip hose and reel pods to allow refueling of Navy and Allied aircraft. (Boeing Photo)

The hose and reel KC-135 is shown refueling a pair of F/A-18 fighters of the U.S. Navy. The Hornet is used internationally by Canada, Australia, and others. (Boeing Photo)

Lockheed Advanced Tanker/transport Concepts

Since it is felt that because of economic restrictions there will be no one-for-one replacement of the KC-135 fleet in the future, Lockheed has investigated an advanced "Box Wing" tanker with a significantly greater fuel transfer capability.

Note the revolutionary configuration that features a pair of forward-mounted swept wings that are joined at their tips by another set of top mounted rear wings. The concept shows a pair of fighters being refueled by a pair of hose-and-drogue systems. In addition to its significant refueling capabilities, the system would also have a significant cargo-carrying capability.

US/European Joint Tanker Proposals
Raytheon/Airbus Industrie MultiRole Tanker/Transport (MRTT) proposal

In 1999, Raytheon (USA) and Airbus Industrie signed an agreement to cooperate on developing the MRTT Tanker/Transport concept. They will use the A310-300 as the base aircraft. This could evolve into the replacement for both the VC-10 and L-1011 tankers. The proposal will bid for the Future Strategic Tanker Aircraft (FSTA) program.

Lockheed Martin/Aerospatiale Matra proposal

These two companies joined forces in the late 1990s to design a new tanker concept. The consortium could also bid on the FSTA program and possible future consideration for a USAF KC-135 replacement.

Foreign Modern Refueling Systems

The United States is by no means unique in the use of tanker aircraft. What follows are examples of what other nations are doing (and have done in the recent past) in aerial refueling operations. The scene changes quickly, but it is safe to say that nations expecting their aircraft to operate in modes of either extended range or time must rely upon aerial refueling capabilities of their own tankers or those of its allies.

Following are the countries listed alphabetically with their tanker inventories. Note that a majority are U.S. tankers or modifications of American tankers/transports.

ARGENTINA
Lockheed KC-130—two tankers

AUSTRALIA
Boeing 707-338C—Modified by Israel Aircraft Industries and Hawker de Havilland, they include four tanker-transport combination aircraft for No. 33 Squadron based at Richmond. Structurally strengthened, the 707s feature a centerline boom refueling system, as well as a Flight Refueling MK-32B hose and drogue pod at each wingtip. They can carry a maximum transferable fuel load of 190,000 lbs. of fuel (about 28,350 U.S. gallons) for support of F-111 and F/A-18 aircraft.

BRAZIL
Lockheed KC-130—Two tankers

CANADA
Boeing 707—two that were modified to support CF-5s originally and now support CF-18s.

CHILE
Boeing 707-320—Carry wing pod-mounted hose and reel systems

COLOMBIA
Boeing 707-320—Carry hose and reel pod

FRANCE
Intertechniqe Transail—Hose and drogue systems on 15 aircraft

Lockheed L1011—Employed in tanker form with hose and reel systems

Royal Australian Air Force 707 two-point hose and drogue tanker refuels an F/A-18 fighter in 1990. (RAAF Photo)

A French Tristar tanker is shown refueling an AWACS 707.

The first British jet tanker was the Canberra bomber. The basket is shown being deployed from the bomb bay. (RAF Photo)

A Tornado F-3 bomber is shown being refueled by a French Tristar hose and drogue tanker.

Another early British tanker was the Valiant, which also deployed its refueling drogue from the bomb bay.

Boeing KC-135FR—Used for refueling nuclear-capable Mirage IVAs. The aircraft are fitted with a rear fuselage, centerline boom equipped with a hose and drogue, as well as near-wingtip-mounted FRL refueling pods. Range is nearly 3,400 miles.

Dassault Etendard IV—A dual-purpose reconnaissance/tanker which carries a Douglas buddy-pack, hose-reel unit centerline under the fuselage.

This British Royal Navy Buccaneer fighter, equipped with a buddy refueling pod, shows the location of the buddy pod on the underside of the starboard wing.

Two Victors from the 55 Squadron are shown in a refueling operation. The Victor receptacle is located directly above the canopy. (RAF Photo)

This Vulcan B2 bomber was converted into a tanker during the Falklands War. It is shown here refueling a Buccaneer fighter.

This RAF KC-135FR tanker carries new Mk.32B pods located on the wingtips in addition to the boom system. (Boeing Photo)

GERMANY

Boeing 707—Four converted to tankers/transports from VIP and support flight carriers.

Panavia Tornado—Equipped with fuselage-mounted buddy-store pod, designated Model 28-300—a hose and reel system built by Sargent-Fletcher Co.

GREAT BRITAIN

Lockheed L-1011-500—Formerly operated by British Airways and Pan American World Airways. To support long range operations, these aircraft were converted to strategic and tactical support aircraft having tanker capabilities. The fuel capacity was raised to a maximum of 312,000 lb. There are three hose and drogue systems, one centerline and two podded, underwing systems. The modified L-1011s themselves can be refueled. Nine exist in 1999, with replacement starting in 2011.

Vickers VC10-1101—Designated Mk.2s and Mk.3s. Technically tanker-transports, these aircraft have five additional fuel tanks in the fuselage. The Mk.2 can carry

18 passengers; the Mk.3, 17. Each has a centerline hose and drogue, as well as podded systems—MK32Bs—built by Flight Refuelling Ltd., on each wing. 24 exist in 1999, with replacement starting in 2007.

Lockheed KC-130J—Deliveries began at end of 1998. Tankers are equipped with refueling booms.

Lockheed KC-130E—Six tankers

English Electric Canberra B. Mk 16—Twin-engine bomber converted for several uses. As tankers, they were configured with hose and reel pods. Sadly, many have been scrapped; a few survive in both public and private museums.

Handley Page Victor BK.Mk 1—Four engine medium bomber modified as a two or three-point tanker (retired)

Avro Vulcan—Originally a four-engined bomber, towards the end of its career some were modified—in 51 days from request to completion—as "K" aircraft. The modifications included a boxy pod which housed hose and

The earlier KC-135s used by the British are identical to those used by the USAF. Note the lack of turbofan engines which were later added to this model. (Boeing Photos)

This Iranian KC707, equipped with turbofan engines, carries a boom and two wingtip drogue systems. (Kalt Collection)

reel systems mounted under the tail cone; the bomb bay was filled with three extra fuel tanks. The tanker configuration was called for during the war in the Falklands in 1982. One of the most strikingly distinctive bombers ever built, the delta-wing aircraft was retired in 1984.

Vickers Valiant—First of the three "V" bombers (Valiant, Victor, and Vulcan). The first prototype flew in May 1951, and a number of these aircraft were configured as tankers with hose and reel assemblies in their bomb bays in 1960. In 1962 the Valiants were assigned low-altitude penetration roles. The resulting turbulence caused wing

An Israeli C-130, modified with BEDEC hose and reel pod system. (BEDEC Photo)

A BEDEC C-130 is shown refueling a pair of helicopters from its hose and drogue systems. (BEDEC Photo)

failures, and the entire Valiant force was grounded in December 1964. Victors were then immediately pressed into tanker service.

Hawker Siddeley Buccaneer—FRL pod-equipped for buddy system.

Supermarine Scimitar—Replaced by the Buccaneer; used for some time as a "buddy" tanker for the Buccaneer (retired).

Hawker Siddely Argosy C.Mk 1—Similar to the C-119 "Flying Boxcar" in design, the Argosy was a transport

This BEDEC C-130 refuels a pair of F-4 Phantoms while a pair of A-4 Skyhawks await their turn. (BEDEC Photo)

An IAF F-16 takes fuel off the boom while a pair of A-4s recieve their fuel from wingtip drogue systems. (BEDEC Photo)

that could be modified for tanker duties with a hose and reel system mounted above the flight deck on the starboard side (retired).

Panavia Tornado—Carries fuselage-mounted buddy store.
Note: As of this writing, the British are looking to replace their VC-10s and TriStars. Candidates include the Airbus A310-300, a version of Boeing's 767-300ER, and possibly a 757 derivative.

INDONESIA
Lockheed KC-130B—Two tankers

IRAN
Boeing 747-100—Three "pre-revolution" aircraft converted to flight refueling

Boeing 707-320—Four tankers

ISRAEL
Boeing 707—Six modified by Israeli Aircraft Industries Ltd. BEDEK Aviation Group for aerial refueling missions. These tankers can transfer up to 123,190 lbs. of fuel 1,150 miles from base. Equipped with a boom and

wingtip-mounted hose and reel pods. Maximum cruising speed at 25,000 feet altitude is 605 mph. Their range with an 88,000 payload is 3,625 miles, or with maximum fuel is 5,755 miles.

Lockheed KC-130—Three KC-130H tankers, built by Lockheed. With their maximum fuel load, including external tanks, they can fly 4,894 miles. Used for refueling both fixed wing and rotary wing aircraft.

Douglas A-4—Fuselage-mounted Sargent-Fletcher Buddy store.

ITALY
Boeing 707—Four ex-airliners converted to tankers

Panavia Tornado—Fuselage-mounted buddy store

JAPAN
Boeing 767T/T—A –300 derivative. A cooperative arrangement with Kawasaki, the work included equipping

OPPOSITE: A cutaway of the Israeli KC-707 shows the retention of seats from a former airliner configuration. Note the two wingtip pods and boom, which is remotely controlled by an enclosed control panel. (BEDEC Photo)

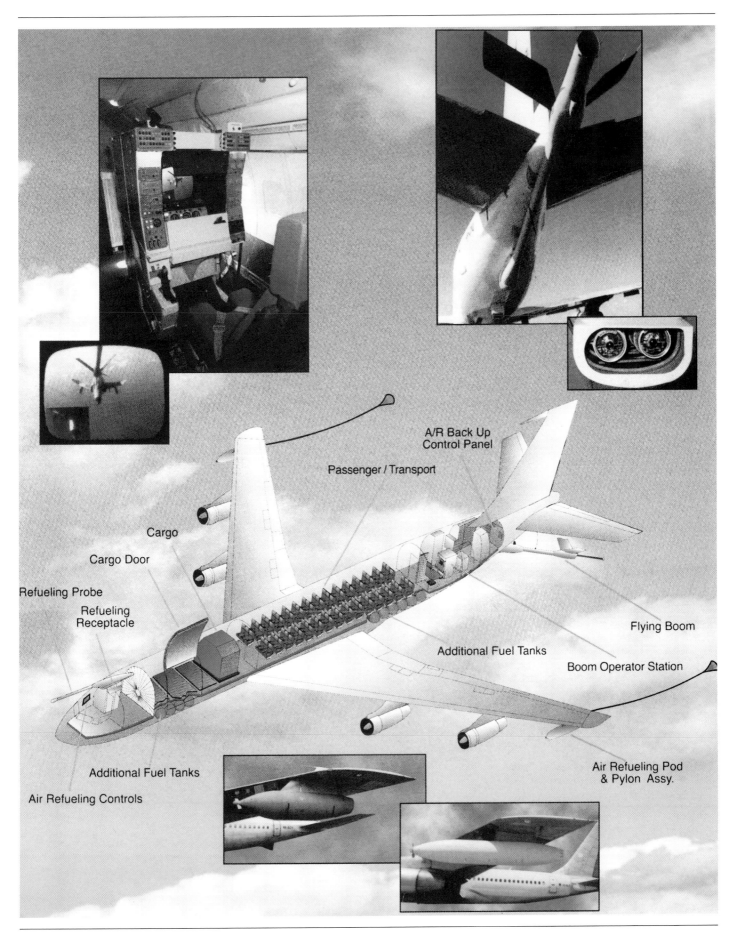

A/R Back Up Control Panel

Passenger / Transport

Cargo

Cargo Door

Refueling Probe

Refueling Receptacle

Flying Boom

Additional Fuel Tanks

Boom Operator Station

Additional Fuel Tanks

Air Refueling Controls

Air Refueling Pod & Pylon Assy.

the aircraft with a refueling boom, extra tanks, and other equipment, including two underwing pods. The boom is remotely controlled from the cabin and assisted by closed circuit TV.

MALAYSIA
Douglas A-4—Fuselage-mounted buddy store

MOROCCO
Boeing 707—The Royal Moroccan Air Force has a short-fuselage –178 tanker, converted by AMIN (Aéro Maroc Industrie).

Lockheed KC-130H—Two hose-and-reel tankers

PERU
Boeing 707-320—hose and drogue pods

RUSSIA
Ilyushin Il-78M—Three-point tanker developed in the late 1970s to replace modified Myasishchev 3MS2s and 3MN2s. The first operational Il-78M had just one refueling pod under fuselage. Initially, storage tanks in the cargo hold could be removed so that the aircraft could be used as a transport; the current version has fixed tanks. The Il-78M can refuel aircraft on the ground by using conventional hoses. It uses UPAZ-1-A Sakhalin refueling pods—hose and reel systems—under outer wings and on the port side of the rear fuselage. The rear turret is used as a flight refueling observation station. It has a maximum takeoff weight of 462,965 lbs.

Myasishchev M-3—40 strategic bombers modified into probe and drogue tankers.

Tupolev 95—Mid-1950s-era bomber, some of which were modified as tankers. Both the Tu-95 and the anti-submarine version designated the Tu-142 had four turboprop engines, each having two, counter rotating sets of four blades. These powerplants were reportedly so loud that F-16 pilots could hear them as they "escorted" the bombers away from allied areas (retired).

This Saudi F-5 is shown being refueled in a 1965 operation. Note the fixed probe located to the right front of the cockpit. (USAF Photo)

SAUDI ARABIA

Boeing 707-320—Eight tankers designated KE-3A are assigned to No. 18 Squadron, Royal Saudi Air Force.

Lockheed KC-130H—Seven tankers

SINGAPORE

Boeing KC-135R - In late 1999, Boeing delivered the first of four re-engined Stratotankers. Originally a KC-135A, this aircraft features reliable, fuel-efficient CFM56 engines and a multipoint refueling system—wing-mounted drogue pods a and centerline boom—developed by Boeing Aerospace Support's Military Programs division, Wichita, Kansas

Lockheed KC-130B Hercules—Four tankers

Douglas A-4 Skyhawk—Fuselage-mounted Buddy Store

SPAIN

Boeing 707—Three at Torrejón: two to refuel EF-18 Hornets, and one, with a secondary tanker role, equipped with signals intelligence equipment installed by the Israelis.